Conversations About The Environment

Conversations About

THE ENVIRONMENT

Edited by Howard Burton

Ideas Roadshow
INTELLIGENT. INQUISITIVE. INTERNATIONAL.

Ideas Roadshow conversations present a wealth of candid insights from some of the world's leading experts, generated through a focused yet informal setting. They are explicitly designed to give non-specialists a uniquely accessible window into frontline research and scholarship that wouldn't otherwise be encountered through standard lectures and textbooks.

Over 100 Ideas Roadshow conversations have been held since our debut in 2012, covering a wide array of topics across the arts and sciences.

All Ideas Roadshow conversations are available both as part of a collection or as an individual eBook.

See www.ideasroadshow.com for a full listing of all titles.

Edited, with preface and all introductions written by Howard Burton.

All *Ideas Roadshow Conversations* use Canadian spelling.

Contents

TEXTUAL NOTE 8

PREFACE 9

SOLAR IMPACT
CLIMATE AND THE SUN
A CONVERSATION WITH JOANNA HAIGH

Introduction... 17
I. Meteorological Beginnings................................. 22
II. Science and Gender...................................... 29
III. A Curious Correspondence.............................. 32
IV. Considering the Earth.................................. 36
V. Considering the Sun 40
VI. The Big Picture.. 49
VII. Examining the Details................................. 56
VIII. Getting The Word Out................................. 63
IX. Public Policy.. 69
X. Final Thoughts... 72
Continuing the Conversation............................... 76

OCEAN ENLIGHTENMENT
A CONVERSATION WITH EDIE WIDDER

Introduction... 79
I. Bright Beginnings 84
II. Bioluminescence.. 90
III. The Eye-in-the-Sea.................................... 97
IV. Grappling with a Coastline Crisis..................... 104
V. Kids to the Rescue..................................... 110
VI. Existential Challenges................................ 116
Continuing the Conversation.............................. 120

CORAL REEFS
SCIENCE AND SURVIVAL
A CONVERSATION WITH CHARLES SHEPPARD

Introduction..123
I. Watery Beginnings 128
II. Building A Reef......................................135
III. Gratuitously Unsustainable 144
IV. Towards Progress? 161
V. Climate Change......................................167
VI. What To Do?..170
Continuing the Conversation...................178

SAVING THE WORLD AT BUSINESS SCHOOL
PART 1
A FIRST CONVERSATION WITH ANDY HOFFMAN

Introduction..181
I. Building a Career....................................186
II. Environmental Evolution 190
III. Beyond Punditry 194
IV. Fostering Debate...................................198
V. American Exceptionalism? 202
VI. Talking the Talk....................................208
VII. Preaching to the Choir? 213
VIII. Energy Renaissance...........................216
IX. Reinventing Sustainability 219
X. Surprising Revolutionaries...................222
XI. Setting Ideals..227
XII. Impact..230
XIII. The Passion Principle 236
Continuing the Conversation...................238

SAVING THE WORLD AT BUSINESS SCHOOL
PART 2
A SECOND CONVERSATION WITH ANDY HOFFMAN

Introduction..241
I. Reprise...246

II. Truth Decay...250
III. The Value of Wisdom 257
IV. Investigating Rewards...263
V. Concrete Opportunities..271
VI. Management as a Calling..277
VII. Opinionated Ignorance...285
VIII. Qualified Optimism..291
IX. Spreading the Word..298
X. Getting Personal..307
XI. Shattered Leadership? 310
Continuing the Conversation...320

Textual Note

The contents of this book are based upon separate filmed conversations with Howard Burton and each of the four featured experts.

Joanna Haigh is Professor Emerita of Atmospheric Physics at Imperial College London and was Co-Director of the Grantham Institute until her retirement in 2019. This conversation occurred on September 16, 2016.

Edie Widder is Founder, CEO and Senior Scientist at Ocean Research & Conservation Association, Inc. (ORCA). This conversation occurred on November 15, 2012.

Charles Sheppard is Professor of Life Sciences at the University of Warwick and has worked extensively for a wide range of UN, governmental and aid agencies in tropical marine and coastal development issues. This conversation occurred on September 19, 2016.

Andy Hoffman is the Holcim (US) Professor of Sustainable Enterprise at the University of Michigan. The first conversation with Andy occurred on May 1, 2013. The second one occurred on December 16, 2020.

Howard Burton is the creator and host of Ideas Roadshow and was Founding Executive Director of Perimeter Institute for Theoretical Physics.

Preface

Ideas Roadshow collection on The Environment is significantly different from virtually all others in that none of the experts involved explicitly set out to become environmentalists.

Joanna Haigh describes how her research career was triggered by student travels after finishing off her undergraduate degree.

"I had a very pleasant three years being an undergraduate physicist—I had a lovely social life—but, at the end of it, I thought physics was pretty dull. So, I had a gap year with my boyfriend—now husband— and went around looking at Middle-Eastern medieval sites.

"And while I was travelling, I saw some amazing meteorological phenomena: enormous thunderstorms in the Mediterranean, fantastic sunrises and sunsets when we were traveling in Morocco, and things like that. I'd always liked meteorology: when I was a girl guide, I did the weather badge and made my own weather gauge and all the rest of it—it was very primitive, of course, but I had always been interested in that sort of thing.

"So I applied to do a Master's degree in meteorology and loved it. After that, I thought, What can I do next, because I want to stay in the field? so I applied to do a doctorate in atmospheric physics, which was the driest sort of physics you can imagine, but that didn't matter because when you know why you're doing something and you're inspired and interested, it makes all the difference. And I've been doing that sort of thing ever since."

Well, "that sort of thing", it should be clarified, involved becoming an internationally-renowned professor of atmospheric physics at Imperial College London and longtime co-director of the Grantham Institute for Climate Change and the Environment.

Edie Widder, meanwhile, simply found herself in the right place at the right time to hone what would later become her scientific specialty: marine bioluminescence.

"When I started my PhD work, I was offered this opportunity to work on a bioluminescent dinoflagellate with a woman named Beazy Sweeney, who was famous in the field of phycology, the study of algae. My major professor was Jim Case, but he and Beazy Sweeney had gotten together and thought up this project of making electro-physiological recordings of bioluminescence from a dinoflagellate that she had in culture.

"I remember having this long conversation with Beazy Sweeney about bioluminescence, where I was sitting there trying to nod and look intelligent while thinking to myself that I don't really know what bioluminescence actually is. I rushed home that night and pulled the Encyclopedia Americana off the shelf—because this was the old days before the internet—and looked up bioluminescence, and it turned out that the entry was written by Beatrice M. Sweeney.

"As it happens, bioluminescence itself was not actually the focus of the research. It was a convenient effector system to record from—I only gradually got more and more interested in this ability of creatures to make light.

"Jim was brilliant at writing proposals, and he wrote a successful proposal to the American Department of Defense for an optical multichannel analyzer—which is a fancy way of saying spectrom-eter—that was so super-sensitive it could measure bioluminescence spectra, the colour of the light.

"He got the equipment and it was just sitting there on the bench in the lab. I've always been kind of a gadget freak, so I couldn't quite leave it alone. I just kept fiddling with it until I pretty well became the lab expert on it. Then one day he turned to me and said: 'Well, now that you know how to make this thing work, we need to start sending you to sea to measure all these animals in the ocean that make light that nobody has ever been able to measure before.' Suddenly I was

a seagoing marine biologist—which is what I'd always wanted to be—but never thought anybody actually gets to be; and I loved it."

Several decades of highly successful and innovative research later, armed with a prestigious MacArthur "genius" Fellowship, Edie set up ORCA—Ocean Research & Conservation Authority in Fort Pierce, Florida, teaching today's youth how to harness the power of biolu-minescence to help protect coastlines.

For **Charles Sheppard**, on the other hand, his youthful love of the sea pulled him, eventually, towards becoming an internationally renowned expert on coral reefs:

"I was born and brought up in Singapore, which was more marine orientated in many ways. I was offered a PhD after graduating in medical research—I was doing a medical physiology and pharma-cology degree—but I began scuba diving as a hobby and began to wonder, What's all this around me? This is really cool stuff.

"And I was very, very fortunate. After getting my undergraduate degree, I turned down that offer and instead took up an offer for a doctorate at Durham with someone called Professor David Bellamy, who offered me many wonderful opportunities. My PhD was about impacts around the UK: we were measuring ecosystem changes as you went into pollution gradients, into polluted areas. Then I got the opportunity to work on reefs, and I had a postdoc in Australia, where it was pure ecology. It was at the Australian Institute of Marine Science and was the best sort of opportunity you could have. They essentially said, 'We're going to pay you for two years, you can do what you like.'

"So, that was pure ecology, but after that I began looking at reefs, focusing on the impacts and monitoring—not only reefs, but mostly reefs—in the Mediterranean and other areas."

Lastly, for **Andy Hoffman**, an environmental awakening was trig-gered by external circumstances.

"I had decided to pursue a chemical engineering degree without too much reflection: I liked chemistry, I liked math. I put as much thought into that as any 18 year old would. And then Love Canal happened when I was an undergraduate. I thought to myself, That's something I can use my chemical engineering training for, to make sure that sort of thing never happens again.

"So I minored in environmental engineering, which at that time was just waste-water engineering. It wasn't focused on pollution, or anything like that. My first job was with the EPA (The Environmental Protection Agency), and I hated it. I worked there for two years and just felt like I was making paper.

"I helped a friend build a deck at the time and got a charge out of it, so I started scanning the classified ads in the newspapers and eventually got a job as a carpenter in Nantucket. I did that for 5 years and then decided to go back to graduate school for construction management. Environmental issues got exciting then, since businesses started doing it because they wanted to. When I was working at the EPA, I was just a policeman: it was just a pain when I showed up and ruined people's days. But now it was strategic. I was offered the chance to do a PhD at MIT and took it.

"There was a lot of activity at MIT at the time. John Ehrenfeld had just started an initiative in business and the environment there, and there was a critical mass of students. It was a very exciting time, right at the beginning when this was all brand new. It was focused on the idea of trying to focus on positive change, rather than mere negative enforcement. When companies started to see that there was a connection between their strategy and their ultimate interest in protecting the environment, that's when it got really exciting."

The fact that these four people, each with widely different backgrounds and interests can now be coherently bundled together into a passionate and eloquent group of like-minded experts speaks volumes about the pressing nature of the environmental challenges we face. And it also, needless to say, puts paid to those who steadfastly maintain that "the environmental lobby" is a meaningful concept, on par with "the fossil fuel lobby" or "the gun lobby".

It is not. These days, being "an environmentalist" simply means being aware of the enormous, deadly serious, and often wholly preventable destruction going on around us.

Solar Impact

Climate and the Sun

A conversation with Joanna Haigh

Introduction

Confronting Complexity

Meteorology is a hard field.

To understand what's happening with our climate you need to understand a wide smattering of physics, from mechanics to electrodynamics to thermodynamics, to get a general picture of the underlying processes at play, many of which interact with each other in far from straightforward ways. But that's only the beginning.

Once you get to the stage of building models, their success will naturally depend on a clear understanding of what is happening both now and in the past, which in turn necessitates the incorporation of a staggering array of meticulous observations of the atmosphere, oceans and ice cores over a long period of time.

Of course, in order to somehow process all this data, you need nothing less than a mind-boggling amount of computing power; but even once you manage to get a hold of that you'll find yourself suddenly facing the aptly named complexity framework of nonlinear dynamics where each run-through will be subjected to the so-called "butterfly effect", ensuring that you perform at least a few thousand separate simulations before you can begin to have real confidence in the value of the outputs.

Hopeless, then?

Far from it.

Joanna Haigh, Emerita Professor of Atmospheric Physics and former Co-Director of the Grantham Institute at Imperial College London was studying the complex interaction between the physics of the stratosphere and the chemistry in the ozone layer when she came across a claim promoting a particularly intriguing correspondence.

> *"There was a paper published by somebody who showed these graphs; one was the temperature in the Northern Hemisphere and the other, a measure of solar variability—it wasn't actually solar irradiance, it was something to do with the sunspot cycles—and in these two graphs the lines almost precisely overlaid each other. The article was in Science, a very reputable journal, and I thought, **That's interesting—there must be something in that**.*
>
> *This was at a time when people were starting to take an increased interest in climate change, and there were people who were saying, "**Well, it's all due to the sun**". And if you looked at these graphs, that's exactly what you might think. So I said to myself, "**I'll take a closer look at that**."*

As you might imagine by now, taking "a closer look" was hardly straightforward. First you have to understand what solar variability really is and what is producing it, and then you have to understand how the authors of the paper had produced their results.

So it was complicated. But eminently feasible.

> *"It became clear that the people who had analyzed the data had—well, let's just say that their analysis techniques left a little bit to be desired. In fact, if you do it properly, solar variability doesn't go up at the same rate as the temperature, so actually, it was quite flawed.*
>
> *"However, that paper was used—quite shamelessly and for many years thereafter—by people who wanted to say that climate change was due to the sun. It's actually only stopped recently, so it had a big impact."*

Welcome to the world of highly politicized scientific research. Yet another complication.

Meanwhile Joanna, having turned her attention to the question of solar variability, began to look more carefully and rigorously at what its precise effects might be.

*"The radiation that is important in the stratosphere for ozone and so forth is ultraviolet, which is why my ears pricked up when I initially heard about this business of solar variability, because the UV variation is much larger—well, it's only a few percent, but that's much larger than 0.1% for the total radiation—which made me think, **Well, that could do something**. So, I started looking at how changes in the UV spectrum could influence the climate.*

*Not surprisingly, what happens is that the temperature in the stratosphere goes up and down quite a lot—by a few degrees—and that's being measured by satellites and it's fairly well understood in general. So, you might say, "**Well, changes in the stratosphere, that's all very interesting for people working on stratospheric chemistry and the ozone layer, but does it have any effect on us living on the earth's surface?**" That's what I've spent some time working on."*

Meanwhile, of course, the question of climate change became increasingly present in the public consciousness. *Was it really happening? To what extent? How confident could we be that the causes were man-made?*

And once more, a common argument invoked by climate-change sceptics was one of complexity. There was, they claimed, simply no way of knowing for sure what was going on. There were all these different models—some people said this and others said that and nobody could know for sure what was actually going on.

But it turns out that's not actually true. A very strong scientific consensus has emerged that there's no way to explain the increase in global temperatures without anthropocentric factors being taken into account.

"You could perhaps get away with it until about 1960 or something, but you cannot get global warming without increased greenhouse gases or some magical factor that nobody's thought of. And I think it's unlikely to be that because if we do the physics and we put all the other factors that we know about into the big models, we can reproduce the temperature change fairly well."

So yes, it's complicated. Very complicated, even. But shouldn't that be a cause for celebration that we've actually been able to truly figure some things out?

You'd certain think so, particularly when such things have a tremendous impact on so many people's lives around the planet.

So why are people so slow to take meaningful global action? Why do so many climate-change sceptics exist despite all the strong scientific evidence to the contrary? Why are so many convinced that global warming has nothing to do with human activity?

Well, some questions might well actually be impossible to answer.

But climate change isn't one of them.

The Conversation

I. Meteorological Beginnings

Joanna finds her niche

HB: I'd like to talk a little bit about how you got into science and atmospheric science in particular. How did that all begin for you?

JH: Well, when I was at school, I was quite good at maths and science. I didn't have any particular leanings in any direction but I had a really brilliant, inspirational physics teacher; and so, when I applied to university, I applied to do physics. I had a very pleasant three years being an undergraduate physicist—I had a lovely social life—but, at the end of it, I thought physics was pretty dull. So, I had a gap year with my boyfriend—now husband—and went around looking at Middle-Eastern medieval sites.

HB: Where did you go in particular?

JH: Turkey and Syria. We tried to go further around but we didn't quite make it around the Mediterranean that way—we had to go around the other way, but that's a long story. And while I was travelling, I decided I liked meteorology. I'd always liked meteorology, actually: when I was a girl guide, I did the weather badge and made my own weather gauge and all the rest of it—it was very primitive, of course, but I had always been interested in that sort of thing.

So I applied to do a Master's degree in meteorology and loved it. After that, I thought, *What can I do next, because I want to stay in the field?* so I applied to do a doctorate in atmospheric physics, which was the driest sort of physics you can imagine, but that didn't matter because when you know *why* you're doing something and you're inspired and interested, it makes all the difference. And I've been doing that sort of thing ever since.

HB: You mentioned your inspirational and outstanding high school teacher. Was this a he or a she?

JH: It was a he. Now, that's funny, because it was in an all-girls school and—the science teaching was actually very good across the board—he was the first male teacher in the school. This is back in the 1960s or '70s—poor guy; young teacher coming into the world of silly, giggling, 13-year-old girls, you can only imagine—but he was really inspiring and taught physics in a way that we hadn't had before. Suddenly the lab experiments were no longer smelly and boring—they were really interesting.

HB: So, it was more the experimental side or was it also the theoretical structure—or perhaps the combination?

JH: It was the whole structure: explaining why you do things and what it means—not just F=ma, but what it actually meant in the real world. As it happens, he died quite recently, most unfortunately.

HB: I imagine he might have inspired quite a few people to go on to do physics.

JH: He did. I mean, this was an all-girls school, as I say, a state school, and it never had a very big physics class doing A level. You always had a few going up to what was then called O level at age 16 or so, because if you wanted to do medicine—and many of the girls wanted to do medicine—you had to have physics O level, but very few took physics A Level. But in my year I think there were six of us, which was quite amazing.

HB: Right—possibly record-setting, I can imagine. I'd also like to ask a follow-up question about your story. You mentioned that you had long liked thinking about the weather—you liked meteorology and you were inspired by physics—but it seems as if, when you were an undergraduate, you didn't put that together, somehow. Why was that, do you think? That's not an accusation, of course, just a question.

JH: No, no, it's not an accusation at all and I'm trying to think of the answer. The answer is twofold, I think. The course was quite dry and there weren't many options available; but also, actually looking back on it, I was immature, having a great time, not concentrating and didn't get as much out of it as I should have done.

HB: Okay—obviously, we're talking about your history and you were there, so I don't pretend to know better—but I have another theory or, at least, I have some smatterings of a potential direction. When I was an undergraduate, which was probably not all that hugely different long ago than when you were an undergraduate, atmospheric science and meteorology and climate science didn't really have a great reputation—it wasn't something that, by and large, the best and the brightest were naturally drawn to.

So, insofar as one is thinking about one's future academically, there was all the sexy stuff about the grand, underlying laws of the universe and so forth, and that might have resulted in meteorology being to some degree regarded as not as attractive as it might otherwise have been. Does that make any sense?

JH: Well, it wasn't really offered. You're absolutely right in your depiction of prevailing attitudes, but it's also true that it wasn't offered: the physics course was very narrow; you had very little choice in what you did and it was basically quantum mechanics or solid-state physics.

But it's true that atmospheric physics has been and still is, to a certain extent, a poor relation—there's a pecking order in physics and the more applied you get, the slightly lower you're viewed. That's still true today—actually, you'd better not put that in because people from Imperial College won't like it.

HB: We're talking about the string theorists and the like, right?

JH: Absolutely.

HB: Oh, they're terrible. We can put that in. *I'll* say they're terrible—I'll take full responsibility for that comment.

JH: There's a certain pecking order in physics, in which the extremely theoretical people are perceived to be, perhaps from the outside as well, higher-ranking in the order than the more applied people. The only thing we have going for us is that we tend to earn more grants.

HB: Of course; and be much more socially relevant too.

JH: Definitely more socially relevant—so there's pros and cons. It suits some people to do one thing and it suits other people to do the other, which is probably rather healthy.

HB: That's very diplomatic. You can tell that you've been the chair of a department.

JH: I learned a lot being chair of the department actually, I learned interesting things about the sciences that I didn't know before.

HB: Such as what, exactly?

JH: Well, what they do in plasma physics and fusion research and all the sort of things about semiconductors, for example. I've been quite outside that sort of thing.

HB: Maybe this is more of a personal statement, and I never claim to have achieved a tremendously high level in any area myself, but certainly, as I get older I get a much deeper appreciation of how interesting things are done in a wide spectrum of fields.

When I was younger, I guess I drank the Kool-Aid or whatever; I had this confidence that I knew precisely what exactly the interesting topics were, and everything else was just dirty, applied, uninteresting, derivative and all these other words that one would throw around.

However, the older I get, the more I realize how incredibly fascinating all sorts of areas can be.

JH: Right, it's like the more you know about, the more amazed you get. Something that's very healthy about science nowadays is I think it's generally understood that doing multidisciplinary or interdisciplinary work is actually useful and to be encouraged, whereas, in the old days, you definitely fell between either going one direction or another—you couldn't go across.

HB: Absolutely. Not too long ago, I had a conversation with Paul Steinhardt (*Indiana Steinhardt and the Quest for Quasicrystals*). I don't know if you know him personally—he's a Princeton cosmologist who has also done seminal work in quasicrystals. Now, relatively speaking, I know a lot about cosmology compared to quasicrystals, which just tells you how little I know about quasicrystals.

Anyway, here's this guy who is one of the primary people involved in the development of the inflationary model of cosmology talking about his experience with quasicrystals and how it began from this fascination with looking at rocks and crystals and associated mathematical patterns.

I mean, the mathematical pattern part I can understand, but when I was younger there would be nothing more boring to me than rocks, with topics like crystallography and spectroscopy running a pretty close second.

But the story starts there, and then continues: it's just unbelievable where it goes—quite frankly, it's one of the most remarkable stories I've heard in my entire life, and it just goes to show you how you can gain entrance to all sorts of fascinating things. I guess that's what this "maturity" thing people have been going on to me about for decades is all about.

Anyway, before we turn to specific details about your research, I'd like to return to your "meteorological epiphany," as it were, because the details are still a bit confusing to me.

This is my understanding of what happened: you obviously had some attraction to and ability in the mathematical sciences, you were inspired by a great teacher in high school, you go to Oxford, you have a good time, you go through some relatively dry and dusty courses, you meet your future husband, you travel around the Middle East and you look at historical sites and then, all of the sudden, you become rekindled with enthusiasm for the weather.

To me, there's still a bit of a disconnect; I get everything up until the trip to the Middle East, but what happened there exactly? Did you go outside during one beautiful night in Syria and suddenly say, *"Oh, weather! Weather is what I want to do!"*?

JH: The interest in weather had always been there—it had never gone away—but there were some amazing meteorological phenomena that I saw while I was travelling.

HB: What were those, exactly?

JH: Enormous thunderstorms in the Mediterranean, fantastic sunrises and sunsets when we were traveling in Morocco, and things like that.

HB: Was your future husband encouraging? Did you talk to him? Were you gushing about the weather? How did that play out then? How does it now?

JH: Yes, we still talk at each other: he goes on about history and I go on about the weather and we sit down and have dinner in between.

HB: Does it ever combine? Do you ever talk about historical patterns of the weather or what the weather was like in the Middle Ages or anything like that?

JH: Well, we almost applied for a job share once and the job share was an advertised position in historical meteorological records and you had to understand the weather and you had to be able to read Latin and we thought, *"Oh good, we can do this,"* but we had no idea we'd only get one salary between us, so that wasn't going to work very well.

HB: How many times does that happen, I wonder, that you have to understand the weather and be able to understand Latin at the same time? It's probably not all that common.

JH: Well, there are people who go in for historical climatology—there's quite an interesting branch of that science where they have to look at old documents. There was a lovely paper written a few years ago by somebody who had been looking at ships going across the Atlantic in the tropics—I forget which time of year it was, but they go farther north or south, depending on the wind directions, and they've got records on how long it took to go. So, by getting all these records together, they were beginning to understand about the wind and how it changed season again, from year to year, and they've collected all this and had to read it all in Spanish.

HB: Interesting—so this was during the Spanish Empire I guess?

JH: Yes, it was in the 1600s, I think.

Questions for Discussion:

1. *What percentage of successful scientists would you guess were strongly influenced by a high school teacher? Do we do a good enough job of identifying and recognizing the role that such teachers play in our society?*

2. *Do you think the hierarchy in physics departments is stronger or roughly equivalent to that of other academic disciplines? If it is somehow different, what do you think might be some of the factors responsible for this?*

II. Science and Gender

Different disciplines, different stories

HB: One last thing before we get into your research. You mentioned going to an all-girls school and how your physics teacher was this very influential and inspiring fellow who went into this all-girls enclave. There are, of course, gender issues in the mathematical sciences and I know you've spoken publicly about those issues from time to time. Are things moving in the right direction, at least? Are we entering a stage when more and more young women are considering careers in the mathematical sciences compared to the way it was when you were in school?

JH: Yes, I think in most schools now, the opportunities are available. Certainly in the case of girls going to do Master's degrees in straight maths, I think it's almost 50/50 now. However, physics and engineering are much worse. At Imperial College we got up to 25% female intake at one stage and we thought we were doing really well, and we worked *quite* hard at it.

We have open days for girls and we try to get female role models, at girls schools particularly, to come to events. I think the issue goes back much further in time, however—if you start talking to girls at age 17, or even 16, you've basically missed your chance—it's much earlier than that when much of the effort needs to be made.

We have to ensure that by the time they're deciding whether they actually like subjects or not and whether it's for them, they can already envision themselves being in that mode. So much of the effort in ensuring that this is possible has to come much earlier, at a more formative period—I would say that it has to go all the way back to primary school.

HB: So, why is there such a disparity? I'm surprised and gratified to hear that, in pure maths, it's now 50/50; and my understanding is that, in other fields such as medicine or law, it's at least 50/50, if not actually a higher percentage of women than men, but why, in physics and engineering, do you think there's still this disparity?

JH: Well, I don't know the answer to that, but there are lots of reasons suggested. Perhaps it's still not very well taught in schools and the girls just think, "*Oh, I don't really fancy that*", and are less confident than the boys.

There's also still a certain, sort of prevailing attitude of, *Boys do engineering and get their hands all dirty and do the mechanics and fiddle around with electronics and girls are okay with computing and so forth, but when it gets to the dirty stuff, they just don't do it so much.*

HB: I see, so physics and engineering—certainly engineering—but even physics in terms of "the real world," as opposed to, say, games and puzzles and–

JH: Definitely with respect to games and puzzles, but when it comes to "the real world" you have to distinguish between the real world when it's "dirty engineering" and the real world where it's actually trying to do good in the real world, like environmental science or biological sciences—the real world that most of us can actually relate to.

I think that's quite why I liked meteorology because you can do the clever physics but you can actually look out the window and see it happening right there around you.

HB: Are there disparities within physics?

JH: Yes, definitely. There are certain areas of physics, which have very few women at all; I'm not going to go into the details, but let's just say that the areas that have more direct military applications generally have far fewer women in them than men.

There are fewer women than men in abstract theoretical physics but there's a certain representation, whereas in some other areas there's almost none.

In our space and atmospheric group, meanwhile, it's pretty much 50% and we're never quite sure whether the girls are choosing us or we're choosing the girls, but I think the girls are choosing us because they like the subject.

HB: Is that fairly common throughout the field? Is Imperial College an anomaly at all, or is that fairly standard?

JH: No, if you go to atmospheric and meteorological conferences nowadays, there's a very good female contingent.

Questions for Discussion:

1. What, specifically, should be done to encourage more girls to be interested in science and at what age?

2. Might a lack of women in a subfield of research also indicate a particular set of prevailing attitudes among the men who are there? In other words might a strongly male-dominated subfield also tend to exclude many men of a certain disposition?

III. A Curious Correspondence

Examining the link between temperature and solar variation

HB: I'd like to move to your research now. Let me start by asking you to summarize what I think your average layperson would be curious to know: which is, for you as a specialist who studies the effect of the sun and solar variability on climate, why should we care about that?

JH: So, why should *you* care or why do *I* care? They're not entirely the same question.

HB: Right. Why don't we start with why you care.

JH: Well, I started off from having done a doctorate in the physics of the stratosphere and the chemistry in the ozone layer before the ozone hole was discovered.

We were doing gas-phase chemistry and what would happen to the ozone and what I was doing was looking at how the ozone change in response to chlorofluorocarbons would respond when there was more carbon dioxide in the atmosphere.

The carbon dioxide gives you climate change and temperature change, and the chlorofluorocarbons give you a chemical change, and what I was doing was fitting the two together to see what happens. That's quite an interesting scientific question that involves radiative transfer calculations.

So I was doing the radiative transfer, and part of that is doing solar radiation, and I just got interested in the physics of the solar radiation in the atmosphere. I came at it from that perspective.

Then, sometime in the 1990s, there was a paper published by somebody who showed these graphs; one was the temperature in the

Northern Hemisphere and the other, a measure of solar variability—
it wasn't actually solar irradiance, it was something to do with the
sunspot cycles—and in these two graphs the lines almost precisely
overlaid each other. The article was in *Science*, a very reputable journal, and I thought, "***That's*** *interesting—there must be something in
that.*"

This was at a time when people were starting to take an increased
interest in climate change, and there were people who were saying,
"*Well, it's all due to the sun*". And if you looked at these graphs, that's
exactly what you might think. So I said to myself, "*I'll take a closer
look at that*".

HB: So, without knowing anything about this, I could imagine that,
if you were to have these two graphs that would look virtually iden-
tical, with one overlaid on the other, that's highly suggestive, as you
were implying, of common causes, common mechanisms and so forth.
Presumably, this was picked up by other people at the time, as well?

JH: Yes, lots of people got interested in it. After quite a short time, it
became clear that the people who had analyzed the data had—well,
let's just say that their analysis techniques left a little bit to be desired.
In fact, if you do it properly, solar variability *doesn't* go up at the same
rate as the temperature, so actually, it was quite flawed.

It could have been simply a mistake. They were using a smooth-
ing filter, which was averaging over several solar cycles. A solar cycle
is 11 years, so if you average over 5 solar cycles, you need 55 years
of data.

In other words you need 27 years either side of the point you're
at, so if you go any further back than 27 years ago then you need
to predict into the future what the data is and they predicted up to
nearly 1990 or whatever the date of the paper was and they had
used a sort of algorithm that meant that, once they started going
around the corner, they would carry on going around the corner as
it were—so it was a, sort of, self-fulfilling prophecy.

I couldn't possibly say whether they had done that on purpose or
whether it was just a mistake, but I hope it was a mistake. However,

that paper was used—quite shamelessly and for many years there-
after—by people who **wanted** to say that climate change was due
to the sun. It's actually only stopped recently, so it had a big impact.

HB: So, getting back to this distinction between why you would be
interested and why a member of the general public or somebody
else might be interested, you quite clearly explained your personal
circumstances and what drew you into the field.

But as we move to considering the relevant and associated interests of lay people, we naturally focus on climate change in particular and related questions, some of which directly overlap with this
question about the role of the sun. *Is it happening? If it's happening,
what's causing it exactly? Is it just caused by the sun? And if it's not
just caused by the sun, what impact, if any, does the sun have on this
phenomenon?*

JH: Exactly. First of all, when I spoke earlier about a distinction
between my motivations and those of the general public, I didn't
mean to imply some "us and them" division, like, "*I have scientific
motivations and the general public can have their foolish, layperson
interest in the subject*"—I just meant that I had a way into the subject
and now I can see there's a much broader reason for doing it.

As you so clearly explained, you have to understand natural
factors that influence the climate so that you can disentangle those
from other factors, which may be due to human activity. So understanding how much the sun is doing or isn't doing is very important
to distinguish that from carbon dioxide.

There is also the factor of volcanic eruptions: if you've got a huge
volcanic eruption, it will chuck a load of stuff up and the particles
sit in the stratosphere for a couple of years reflecting sunshine back
to space and that cools the climate, so we want to know how much
that is doing as well.

There are other natural phenomena, not so much in terms of
specific forcing effects, but more in terms of natural variability: so
you need to be aware of El Niño and so forth, and there are other
natural variations in the climate that you need to understand so that

you can be very clear about what the human activity is doing and what it isn't doing.

Questions for Discussion:

1. Are you surprised that such a scientific article that was shown to have such a flawed analysis in its calculation made it through the peer-review process? How often do you think something like this happens?

2. How can non-experts know if those with a particular policy agenda are using flawed science to support their beliefs? Do scientists have a moral obligation to publicly denounce cases where people are relying on scientific arguments that are known to be invalid?

IV. Considering the Earth

A changing orbit and changing tilt

HB: One of the things that I found quite interesting in the briefing paper that you pointed me to from the Grantham Institute is how one can clearly separate the sun's influence on the earth, broadly defined, into factors that have to do with the variability of the sun itself—due to stellar processes and what have you—and those associated with the earth, related to the orbit of the earth, its tilt and so forth.

I think this is an important point to emphasize: that as we consider variation in the amount of radiation that we're receiving from the sun, it's not all about the sun itself. There are all these things that are happening with respect to the earth's orbit and tilt that would change the amount of radiation that would hit the earth, even if the sun's radiation was constant.

JH: Yes, exactly that. There is some misunderstanding, even among experts, as to quite what you're studying when you're looking back in history at solar radiation and climate—you have to be very clear what it is you're actually using as your indicator of solar activity.

You've expressed it very well: the climate is driven by energy from the sun, fundamentally, and that's driving all the weather patterns and the winds and everything; there's almost nothing coming from below. So we need to know what the sun is doing or how much energy is getting in from the sun in order to understand how climate works.

The energy that comes in from the sun is determined by several factors: how much energy is coming out of the sun and where the sun is in relation to the earth and the earth's tilt and position and so forth in relation to the sun. So we need to be careful in distinguishing those different factors.

HB: OK, so I have a couple questions about this. If memory serves, there are three basic effects that are at play if one just considers the earth and how it's moving around and changing.

The first point is that the eccentricity of the orbit is changing a little bit, and I think, *Okay, there are probably tidal forces from other planets and maybe some effects from the moon in various positions with respect to them or something.* So, that's one factor—and clearly as the orbit becomes more or less elliptical and the earth correspondingly moves closer or father away from the sun, the amount of radiation received is greater or lesser. That, I get, or at least I can wave my hands around and pretend that I get.

Then, there's the fact that the earth is basically this spinning top with a particular axis and, as a result of that, like most spinning tops, its axis of rotation will change—it will precess—and such a precession will wreak havoc with our seasons, but presumably this is happening on some long timescale, so fine.

But I also read that the obliquity—the actual tilt of the earth's axis—also changes, and that was surprising to me. Maybe I'm missing something obvious from some undergraduate-level classical mechanics course, but I was surprised by that. How does that happen and how can we be certain that it's actually happening? Apparently we knew that was at 21° 20,000 years ago and 28° at some other point—I can't remember what the numbers were exactly at the moment—but this seems a bit mysterious to me: what's going on and how do we know that that's the case?

JH: What's going on is a combination of influences of all the different gravitational forces from all the different planets over long timescales and how we know is largely based on mathematical models. I'm not an expert on that at all, but there are people who put these many-body problems into various mathematical models and can actually predict each of these three parameters and what they do through time.

It's also very important because, as you say, if you've got an axis that isn't tilting, then you don't get the seasons and you get the same amount of radiation all the way through the year.

And one of the very clear things that comes out of the records of temperature is a quasi-hundred-thousand-year, up and down, change in temperature and we think that's due to changes in the eccentricity of the orbit on a hundred-thousand-year timescale.

HB: So my sense is that these three factors—the change in eccentricity of the orbit, the precession of the orbit and the change in the actual tilt—all happen with some well-defined periodicity, which is quite long—

JH: Yes: a hundred thousand years for the eccentricity.

HB: Right, and I seem to recall that the other effects are also periodic on something like a thirty–forty-thousand year cycle or something.

So, my very hand-wavy reading of this was something like, "*Okay, there are these three independently periodic functions that occur if you look at the relevant Newtonian mechanics of the situation, and when you put them all together sometimes they align and reinforce one another and sometimes they cancel each other out; and when they reinforce one another they presumably cause these massive climatic shifts like ice ages.*" Is it something like that?

JH: Yes, it's exactly that. This is the thing about ice ages: if you have quite a big tilt, then it takes some time for the ice to build up in winter and to melt in the summer and it can be that, if you go through a big tilt and a big eccentricity, there's not enough time for it to melt in the summer before you get to the next freezing and that's when you get the build-up of an ice age.

So, it's a combination of the tilt and the eccentricity of the orbit that's giving you the ice ages.

HB: I see, sort of like the perfect storm, as it were.

JH: Yes, exactly; and they're going on the hundred-thousand-year timescale.

Questions for Discussion:

1. *Is it possible to evaluate the degree to which humans are influencing our climate without a detailed understanding of natural climate variations?*

2. *How does separating natural effects into "earth-related" and "sun-related" make it easier to understand what is actually happening? To what extent is this a standard treatment in science? Are there times when it doesn't work?*

V. Considering the Sun

Looking at the solar cycle

HB: Let's move now to considering the sun itself. We discussed the natural changes that occur to the amount of radiation the earth receives from the sun solely due to changes in its own orbit and axis, but the sun is also changing, of course.

Earlier you mentioned the solar cycle and I'd like you to talk a little bit more about that. My understanding is that there is a cyclical pattern in terms of the sun's activity of roughly 11 years, or something like that, right? This is the solar cycle, which people have known about for a couple hundred years or so because it's linked to sunspots, right?

JH: Well, they've known about sunspots for much longer than that. The ancient Chinese recorded observations of sunspots, so they've been known about since ancient times but it's only since about the last 200 years or so that we've known about the 11-year sunspot cycle. However, people were recording the observations of sunspots much longer ago than that: in the Paris observatory, they have very clear records of sunspot counting since roughly the 1500s.

HB: But they didn't notice a cycle or any pattern to it?

JH: Well, it comes and goes. In the 17th century, there was a thing called the Maunder Minimum, in which the sunspots hardly appeared at all and people said, "*Oh, perhaps it's because they just didn't notice them or write them down,*" but that wasn't the case. They knew that there were sunspots and they were looking for them and they weren't happening.

And then in the mid 19th century the German astronomer Samuel Schwabe discovered this 11-year cycle in the sunspots. However, it's only very approximately 11 years—when you plot the graph it looks very clear, but it actually varies between 9 and 13 years and actually, the amplitude of the cycle varies a lot, so you get a small one and then a bigger one, so it's not quite as obvious as you might think if you were just sitting there looking at the data.

HB: So, by the amplitude of the cycle, you presumably mean the total number of actual sunspot detections during the peak period. So how many would that be, exactly?

JH: Well, it depends on how you count them—and that's another issue because they come in a range of sizes. It's just been revised, but there is now an official way of counting sunspots where you count individual, big ones; then you count groups of them; and you multiply the number of groups by 10 and you add that to the number of big ones and that gives you the formal sunspot number. That's the idea anyway.

So in answer to your question of how many, there are a few hundred at a typical peak using that sort of measure—120 or something.

HB: Okay, now the obvious follow-up question is, *"Given that something is happening cyclically in terms of the appearance of these dark spots on the sun, what is going on? What are they anyway?"*

JH: That's an interesting question because, actually, they're not dark at all—they only look dark to us. There's lots of radiation coming out of them but they just look dark because it's much brighter around the outside and they're recessed, so light doesn't escape so well.

If they were just sitting there, without the rest of the sun there, they would still be very bright.

HB: And when you say there are 120 of them or something at the peak, what is the time period that this number corresponds to? Is that daily? Weekly?

JH: That would be an instantaneous number, and they come and go.

HB: So, at any given moment, there are a hundred or more of these things at the top of the cycle?

JH: Yes.

HB: And there's a pattern to the appearance of these spots that's approximately 11 years—as you say, it's roughly 9 to 13 years—but there's a clear periodicity.

JH: Yes; and then you get weaker cycles and stronger cycles and the weaker cycles are also longer ones. When we were talking earlier about that record of solar activity that had been used, what they actually used was the length of the solar cycle to indicate how powerful the sun was, so they had used an inverse scale. What that meant was a long solar cycle meant low solar power and that was why they were having to use very long records to get any measures.

HB: So, what's going on with these things, with these solar cycles? What's causing them?

JH: Well, I'm not a solar physicist, I'm afraid.

HB: You're the closest thing to one that I've ever talked to. Moreover, you're most definitely the specialist in the room at the moment.

JH: I only know in the most hand-waving sort of way, but what happens is that there's little convections. It's like how we have convective clouds on earth but these are convections of magnetic storms on the sun that start at high latitudes and they move, and you get these very complicated magnetic field lines that get twisted around the sun and eventually move inwards towards the equator of the sun. Then, the next lot starts at high latitudes and moves in again. That's how you get the 11-year cycle.

Now, if you asked a solar physicist—and I have been to conferences and listened to them and tried to understand what's going on—they can understand the physics of what's happening much better than I can explain it, but if you ask them *why* it's specifically 11 years they really can't explain it—they have to put parameters into their models and they can't explain where they're coming from.

They look at other stars and they've all got little cycles as well—well, I'm not sure about *all*, but at least most sun-like stars have cycles as well. The same physics is going on, but why it has this particular periodicity, I think, is still not known.

HB: Do the other stars have a similar periodicity?

JH: They vary. There was quite a lot of work in looking at the sun's history and comparing it to sun-like stars to see how its radiance in the past would have varied based on the sunspot numbers; and there's a whole discussion about whether the sun-like stars were really sun-like and that sort of thing, but in short: yes, they do. There's a whole class of stars that are like the sun and have these variations.

HB: So presumably, this is related to the size and age of stars or something?

JH: Yes, something to do with the age of the star causes this effect.

HB: And my understanding is that sunspots are just one manifestation of the solar cycle, but there's also a measurable change in the energy flux reaching the earth, right?

JH: Yes.

HB: So when there are lots of sunspots, there's more energy that's hitting?

JH: Yes. So, if we think about what energy is coming out of the sun, most of the energy is coming out in electromagnetic radiation in visible wavelengths that we can see—there's a whole great spectrum of that. There are also a lot of particles that are coming out of the sun, energetic particles—electrons and protons are coming out and they're coming out in flares and storms and things like that.

So, if we're looking at how the sun varies, it's on a huge range of timescales from minutes—where you get these eruptions and storms and plasma ejections and things happening on the scale of minutes—through to decades.

There's also the rotation of the sun—that's a geometric factor actually, not a variation factor—but sitting on the earth and looking at the energy coming from the sun, it varies with the 27-day rotation period, a bit like a lighthouse. So, if you were to suppose the sun was bright on one side and dark on the other, you could imagine it flashing every 27 days, so that's an effect you have to consider as well.

You can actually see the sun rotating easily because of the sunspots—you can go to the SOHO (Solar and Heliospheric Observatory) website, say, and track sunspots moving across the sun: after 13 days it goes from one side to the other.

HB: So, what's the energy differential, roughly, in terms of percentages, as we move through this 27-day period—given that it's obviously not the case that the sun is not bright on one side and dark on the other.

JH: It's of the order of about 0.1% in terms of the total irradiance coming out when there are lots of sunspots around, which is similar to the magnitude of the change between the minimum and maximum of the sunspot cycle—which is not very surprising, of course, because the minimum and maximum of the sunspot cycle corresponds to no sunspots and lots of sunspots respectively.

HB: Right. So another obvious question that somebody might have is, *"Well, okay, I can see that you can count sunspots, but how do I know that that's linked to this 0.1% energy differential? How do I actually measure such a thing?"*

JH: Well, we know that now, but it's only been known relatively recently since we've been able to make precise measurements with satellites.

This was a large open area of scientific research back in the 1930s into the 1950s. People knew that the sunspots were coming and going and they had pretty decent radiometers that could measure irradiance from the instruments on the ground and they were trying to see whether there was more or less irradiance when there were more or less sunspots.

Some were thinking that there would be less radiation when there were more sunspots, because you had black spots blocking out the radiation—a perfectly reasonable hypothesis.

However, they couldn't actually make accurate enough measurements. We're talking about 0.1% here, after all—if you've got an instrument on the ground you haven't got that sensitivity to see that variation with all the atmosphere in the way and the clouds and so forth; and even if the atmosphere is clear, you have turbulence and things going on.

So, it was only when we got satellites sitting outside of the earth's atmosphere with accurate measurements of the radiation that we could conclude with certainty what we now know: that there's more energy coming out of the sun when it's got more spots. And that's because, while there *is* less coming out of the spot area, the sun is more active and there is more coming out of the surrounding areas which more than compensates for the sunspots, so there's a total of more coming out.

HB: Something else that quite interested me were these other indicators, these so-called "proxy indicators" that you talk about.

My understanding is that when the sun is more active, this results in less cosmic rays hitting the earth. Why is that, exactly?

JH: What's happening is, when the sun's more active, it also has a stronger magnetic field and so these particles are steered around the outside, as it were, so they don't reach us as often.

So, as you say, when the sun's more active, you get less galactic cosmic rays coming to us—but it's slightly confusing because the sun itself emits particles, including solar cosmic rays, and there are more of those being produced when the sun's active.

At any rate, we have galactic cosmic rays, which are coming from galaxies and eons away and the incidence of those is fairly constant. However, when the sun is more active, these are steered away by the stronger magnetic field.

HB: And another of these "proxy indicators" is an increase in the likelihood of seeing the Northern Lights, right?

JH: Yes, this happens when these charged particles from the sun are steered around by the earth's magnetic field into the poles, so as the number of those particles increases, the likelihood of this happening also increases.

HB: So, there are all these other tangential or proxy indicators when you can see other things happening because the sun is more active than it was before.

And that's also hugely relevant to being able to answer questions like, "*How do we know what happened 10,000 years ago or 15,000 years ago or 100,000 years ago?*" because some traces of these interactions actually exist in ice cores and so forth.

JH: Yes. Cosmic rays interact with atoms in the atmosphere and in the earth and produce higher levels of isotopes. So, in beryllium and carbon, you get higher isotope levels when there are more cosmic rays and then those atoms will decay back to their normal state over very, very long periods. And you get a higher proportion of those when there have been more cosmic rays.

You can then use the fraction of that isotope to the normal isotope to tell you how active the sun was when those molecules were forming. So, you can look at carbon-14 and you can look at beryllium-10 and both of those things will tell you about cosmic rays that have been influenced by changing solar activity.

HB: Okay, so let me back up for a moment and try to summarize a bit.

We start off from saying that we sometimes get more radiation from the sun than other times. Some of that is because of the earth—the changing shape of our orbit and the changing angle of our axis, both of which we can calculate, or at least model—and some of that is because of the sun. And when it comes to the sun, we can use these proxy indicators, like the relative presence of these isotopes you were just talking about, that give us a clear sense of what happened in the past.

And we're trying to rigorously quantify all of this because we want to understand how the sun influences climate, and to what extent the variation in the energy we are getting from the sun may or may not be responsible for changing temperatures.

So in order to evaluate *that* we need to turn to the historical record and compare variation in solar activity to variations in temperature.

Now, you've described how we determine variation in solar activity through analyzing these isotopes. How do we determine variation in temperatures in the past?

JH: Well, if you've got an ice core, you can use the oxygen isotopes to tell you about the temperature variations. But you have to be a bit careful, because the beryllium that you're measuring to detect solar variation is also affected by temperature, so there's a complication.

HB: But you don't have one, single indicator; so if you take both into account—the carbon and the beryllium—you should be OK.

JH: Exactly that: if you find that the beryllium and the carbon are both doing the same thing you can be fairly sure that it's due to the sun, whereas if you find the carbon is doing something different than the beryllium, then you know that temperature variation is playing a role there.

Questions for Discussion:

1. How do you think the ancient Chinese were able to record observations of sunspots?

2. Are you surprised that, with all her evident knowledge and expertise, Joanna does not consider herself a "solar physicist"? What does this tell you about the degree of specialization of modern science and the amount of understanding that each subdiscipline requires?

3. Do you imagine that all stars have sunspots? Why or why not?

VI. The Big Picture

More than just the sun

HB: Right. As I was reading this, I was struck by the complexity involved.

Even once you get a handle on all of this in one location, you have to then recognize that there are regional differences that might come into play.

And then, there's the fact that some of these variations are frequency-dependent and will result in different effects in different places—a point I'm anxious to talk about later, because I know you've done some seminal work on that.

But however complex it may be in detail, that hardly means that you can't reach some important big-picture conclusions about the factors underlying climate change.

JH: Yes. If you start from the question of, "*Is global surface air temperature changing—the global average?*" then you begin with a very simple perspective in terms of what's called "radiative forcing of climate change."

You get a radiative forcing by changing the sun's radiation; you get a radiative forcing if you increase carbon dioxide or other greenhouse gases; you get a radiative forcing if you put volcanic aerosol up in the atmosphere; you get a radiative forcing if you change the brightness of the surface—and that's a fairly simple concept.

So, we start off with the radiative forcing due to the changes in the total radiation coming out of the sun at any particular time, and we can get a zeroth-order estimate of what its effect on temperature would be and on timescales of say, hundreds of years, it is pretty small.

There's a lot of discussion about this Maunder Minimum in sunspots in the late 17th century as to whether or not that was associated with what's become known as the "Little Ice Age"—a cooler period in temperatures, particularly in the North Atlantic region.

It's possible that it had an effect but it may also be due to the fact that there had been quite a lot of volcanism at that time, so there were lots of volcanic aerosol particles cooling the climate.

So the bottom line is that on those long timescales, just looking at the global average, the sun's effect is very small.

HB: Right. After you do all of these models and all of these calculations and you look at all the possible mechanisms—if you take that all into account, and you're presumably as liberal as you can be with all relevant parameters in your models, you find that there's no way that you can explain the increase in global temperatures without anthropocentric factors being taken into account. Is that a fair assessment?

JH: Absolutely. You could perhaps get away with it until about 1960 or something, but you cannot get global warming without increased greenhouse gases or some magical factor that nobody's thought of. And I think it's unlikely to be that because if we do the physics and we put all the other factors that we know about into the big models, we can reproduce the temperature change fairly well.

HB: Another point to mention here, I think, is that it's not just a question of determining what are the factors responsible for climate change or being able to conclusively demonstrate that those who are claiming that it's all due to a natural fluctuation in energy from the sun are wrong.

Which is to say that even once we recognize that man-made factors are responsible for climate change, a rigorous understanding of the variation of the sun's energy is also necessary for an appreciation of what is going to happen next.

JH: Yes. So as I've just said, the long-term, global average temperature change due to the sun is probably quite small, but if we look in different parts of the globe, we find that there are statistical correlations that seem to be robust with solar activity and local temperature—particularly, say, in the North Atlantic region—you tend to get colder winters when the sun is less active.

That's a nice question to use to try and understand what's going on there; and then we start to look at the factors that will change the regional distribution of winds and temperatures rather than just the global average temperature—which is important because, if you're sitting somewhere in particular on the globe, the global average temperature to use is not, perhaps, of particular interest, and that absolutely applies to climate change as well. I mean, global warming may be one thing, but if you're sitting in a part of Africa where the temperature is going to be rising 10° when it's already 35°, it's much worse. So, the regional effects are important, and not just temperature, but winds, precipitation and all the other things that really affect people rather than just the global average number.

So, it's important to understand what's going on—and, as you suggested, trying to understand whether these effects are actually linearly additive or not is a very interesting question; whether or not you can add some CO_2 to some changes in sun and get the sum of the two parts or whether, if you do them together, you'll get something different, is a very interesting question.

HB: As I said, I look forward to asking more specific questions about your research shortly, but when you talk about non-linearities and CO_2 one of the first things to come into my mind is the greenhouse effect. Maybe you can say a word or two about that.

JH: Well, the greenhouse effect is a wonderful, fundamental part of the climate—if we didn't have the greenhouse effect in the atmosphere, the global average temperature would be about 30° colder than it is, so it would not be the nice, habitable place that we live in.

More than two-thirds of that is due to water vapour in the atmosphere and that's the water vapour that's keeping us warm and comfortable. Naturally occurring carbon dioxide is the next most important one. We understand the physics of that pretty well—it's been known about since Fourier or so, it's fairly well established.

However, you're right to suggest that there are nonlinearities. What happens is that if you warm up the surface of the earth, you will get more water vapour evaporated from the surface. And more water vapour in the atmosphere will give you warmer temperatures than you would have gotten without that subsequent effect, so all those things need to be taken care of.

They're actually the simpler things to work out because it's just fairly simple thermodynamics to do that sort of thing. However, when you start talking about regional meteorology and the sort of chaotic system in which the whole of the earth's climate is in, it's all very complex.

So, you sometimes get people, for example, going back to the global average again, saying things like, "*You tell us that the carbon dioxide is going up and we'll accept that, we can see the graph of it going up and it varies annually but it's pretty much going up year on year, but the temperature isn't going up year on year, it's wobbling around all over the place. Therefore, you're wrong, it's not the CO_2 that's doing it, there's no relationship.*"

There was a recent controversy about what was called a "hiatus in temperature" where the warming didn't go up as much as it had been going for perhaps 15 years or so—it flattened off a bit but is shooting up again now. As a result, there was a lot of discussion about what was going on, because it clearly wasn't a result of the CO_2.

It's like the whole thing is sloshing around in a bucket—you've got the atmosphere, you've got the oceans and everything else swirling around together, so the fact that it's not going smoothly is absolutely not a surprise at all. You need to have a longer-term view in terms of how you study it: what you need to do is run models many times, which is exactly what we are doing with these big climate models that are trying to simulate what's going on across the world.

You always have to start them from some position and some data or some parameters initializing them, but if you just change initialization parameters very slightly, it will evolve in a different way. So, what we have are ensembles of model runs: you start them in a slightly different way and then you follow a trajectory and you get wiggling with y, and a different wiggling with x, and you get a hundred of them.

The point being that, when you've run lots, you get the broader picture. So you can see the climate change emerging from this picture, whereas if you look for year on year, you wouldn't see it nearly so much because it's such a noisy system.

This is an important point that is quite difficult to get over to people when you're trying to explain climate change. You can't predict what is going to happen in July of 2050 but you can predict that, in 2050, it's likely to be so much warmer than it is now, given this range of uncertainty due to the natural complexity of the system.

HB: Right, so there's the question of doing many model runs and there's also improving the resolution of your model—ideally both is what you're aiming for, presumably.

JH: Yes, that's actually interesting because, in terms of weather fore-casting—which, of course, is put in much shorter time periods—we use exactly the same climate models, but the higher resolution you have—better spatial scales and so forth—the better weather fore-casting you can do.

In climate prediction, there's often a feeling that if you get better and better resolution you should get better and better predictions, but it's not entirely obvious that that's going to be the case because you still have to run your ensemble of predictions and it may be that you'd better use your computing time in doing a wider ensemble with more components than one run with very high resolution, though that would still have the natural complexity in it.

HB: To slightly rephrase what you're saying, I'm guessing that many people have probably heard of notions such as of chaos theory—the butterfly effect and so forth—associated with weather forecasting:

this notion that there are instabilities all over the place and you can't predict when they're going to arise.

So one way you can generally try to improve the simulations is to have a really fine-grain filter, increasing your resolution, so that you know every millimetre or whatever it is, of your area, but however fine-grained your filter, you'll always have instabilities everywhere. But as you say, if you increase the ensemble and you run these simulations for 5,000 different possibilities or scenarios and 3,500 of them wind up on one side, that should be a pretty strong indicator.

JH: Yes, that's absolutely the point. In terms of complexity, it's a naturally complex system, and you're not going to make that go away however fine a resolution you're going to get, because that's just part of the whole system. You can reduce numerical instability in your computer models and you can show mathematically that you get instabilities due to the way that you formulated the equations, but you can't get rid of the actual, real, physical things that are going on all the time.

As you say, it's much better to just do lots and lots of runs and get an envelope of variability into the future, so you just get a range of uncertainty but some general pattern will nonetheless emerge.

Questions for Discussion:

1. To what extent is it reasonable to have a strong opinion about the causes of climate change without a genuine understanding of the physics and the relevant models?

2. What, exactly, do Howard and Joanna mean when they talk about "nonlinearities"?

3. What is "the butterfly effect"; and what does Joanna mean, exactly, when she says that "you can show mathematically that you get instabilities due to the way that you formulated the equations, but you can't get rid of the actual, real, physical things that are going on all the time"?

VII. Examining the Details

Recreating the weather, more or less

HB: You spoke earlier about how, notwithstanding the various complexities inherent in these systems, by using general radiative forcing arguments we can confidently conclude that global warming simply **can't** be solely the result of the sun, and we necessarily need to take anthropocentric factors into account.

So that was a very important high-level argument to make, particularly in terms of the awareness of climate change and its causes for the general public, which I hope to speak about later. But now I'd like to wade into some of the details about what we **can** say about our understanding of the climate and how to go about successfully modelling it.

JH: Yes. So, if we say now that we're not going to bother about or we're not going to look at the change in total irradiance, because we've just discussed that that's actually very small and doesn't have much effect, then it's quite interesting to look at the sun's spectrum.

The sun has a spectrum of radiation leaving it, which is peaking in the visible, but it's got wavelengths all the way through from the ultraviolet on one end and infrared in the other. The interesting thing about that is, if you look at the 11-year sunspot cycle, as we said you get a 0.1% variation in the total irradiance, but it's very strongly frequency-dependent: the radiation varies by much larger fractions in the ultraviolet than it does in the visible part of the spectrum.

The visible part is actually controlling the total—because that's where most of the energy is—but the variations in the ultraviolet vary considerably: by the time you get down to 100 nanometres, it's doubling between solar maximum and solar minimum.

That's why I got so interested in this—there was just such interesting physics going on there. There was even some work, which is still being investigated, that there was less radiation coming out in the near-infrared region when the sun is more active than when it's less active—so out of phase. That's still being worked on, so that might not be our exact understanding at the moment, but the point is that it's hugely wavelength-dependent.

And the implications of that are that the sun will affect different parts of the atmosphere differently than the very simplistic idea of just responding to the total amount of energy. So, you asked me before about my career and how it evolved and I told you I was working on the stratosphere, because the radiation that is important in the stratosphere for ozone and so forth is ultraviolet.

And that's why my ears pricked up when I initially heard about this business of solar variability, because the UV variation is much larger—well, it's only a few percent, but that's much larger than 0.1% for the total radiation—which made me think, *Well, that could do something*. So, I started looking at how changes in the UV spectrum could influence the climate.

Not surprisingly, what happens is that the temperature in the stratosphere goes up and down quite a lot—by a few degrees—and that's being measured by satellites and it's fairly well understood in general. So, you might say, *"Well, changes in the stratosphere, that's all very interesting for people working on stratospheric chemistry and the ozone layer, but does it have any effect on us living on the earth's surface?"* That's what I've spent some time working on.

In fact, there's a whole range of people working on how changes in the stratosphere—for many reasons, not just the sun—could be influencing the climate at the surface. The mechanisms that seem to work or that seem realistic are not due directly to radiation—I talked about the heat radiation and the solar radiation, so that's all happening up in the stratosphere—but what happens down below is due to changes in the wind circulations, which occurs as a response to the change in temperature gradients in the stratosphere.

So, if the sun is warming up different parts of the stratosphere more than others you get a temperature gradient, which produces a change in the winds, which affects the circulations down below, and then you can see how you might get an effect at the surface.

We ran some fairly simple modelling studies of that effect and it looked plausibly similar to the observational changes, so we think that might be happening, at least to a certain extent.

HB: Let me ask you more details about these models. How confident are you that they are enabling you to identify these effects?

JH: It's really two questions: how good are the models and how do the mechanisms pan out in the models? Well, to start off with, if somebody doesn't understand what a model is, we should explain that it's not a box with some air in it that I'm staring at on the lab desk trying to simulate weather. It's not a model like that, like a model car. It's a numerical model; it runs on a computer.

They're based on fundamental physics that we understand quite well, so I'll get a bit into details here:

We begin with Newton's second law, $F=ma$, with F being force, m being mass and a being acceleration.

The relevant forces in the atmosphere are gravity, a pressure gradient force—if you have high pressure over here and low pressure over there, that will tend to make the air try to flow from high to low—but it doesn't, in fact, because there's also a Coriolis force that is due to the earth's rotation, and you've also got to take into account viscosity and that sort of thing.

So, you write down all of these forces involved and then you can work out the corresponding acceleration.

Then you've got the continuity equation or the conservation of mass, which basically tells you that what goes in must come out, or else it gets squashed.

Then, you've got thermodynamics: if you add heat to something, it will get hot and expand.

And then, finally, you've got the ideal gas law, which you may remember from school—PV=nRT—that's the relationship between pressure, temperature and density.

So, those are very basic, very well understood physics equations, and we can write them down in a form that can be solved on a computer.

When I say "solved," it's like the simultaneous equations that you did at school, where you have two equations and two variables—x and y, say—so you can work out the answers for both variables.

With these equations I've just described, we actually have six equations and six unknowns—you may have been counting and said, "*Wait a minute, you only have four equations*", but I actually have six because the wind has three components—it can go forwards or backwards or sideways or up and down—so there are three winds, pressure, density and temperature: six equations and six unknowns.

But there's also time variation in those equations, so you have to iterate, and that's where the forcings come in—you have to represent the heating in there and how it varies with time.

Then, you've got all sorts of things like water vapour—which is involved in the heating due to the greenhouse effect, but more importantly, it's involved in the heating due to the fact that it condenses out into clouds. This is where the rather messy "real physics" gets into the models.

So, it *is* pretty basic physics, and we can construct these models in ways that are fairly well-established, but there are these important details that need to be done well. How well are they done? Well, one clear way you can test them is by asking, "*How well can you simulate the present climate?*"

And it absolutely amazes me, still, that we can write down these equations and solve them mathematically, and what we can come up with is something that gives you the northwest trade winds, it gives you the easterlies, it gives you the pressure patterns with the storms going along in mid-latitudes, it gives you rising air in the tropics and deserts in mid-latitudes. It's just wonderful; it's marvellous.

Will it get you the weather precisely right, all the time, in every place? No, but it suggests that the fundamental physics that's in those models is pretty robust.

So then, the question is, To what extent can you take them outside of their comfort zone? You do have to put parameters in to do with the convection and the clouds and all these complicated things—so, *How well you can use those parameterizations for a scenario, which is different from where you started? How good are those simulations when you're doing a climate change scenario?*

In order to probe that you can do something called a "hindcast"—a historical simulation: we'll put in what we know about changes in the sun and volcanoes and all the rest of it and see how well our models can map to the historical climate. And those are not bad at all. I mean they're not perfect and there are certain things that are still difficult, but they're not bad.

HB: Have they gotten appreciably better over the last 10 or 15 years?

JH: They're definitely getting better. If you look at the global average, they're very good now, but if you look at regional patterns, there are some problems, which are related to things like El Niño simulations—we can't precisely date those because the physics of the air-sea interaction is quite complicated: it's generally understood, but not to the detail that we would like.

As I said about predicting future climate, you can't predict precisely on a particular day what it's going to be like, but you can predict typical patterns. It's the same sort of thing for hindcasts.

HB: You mentioned the climate mechanism due to solar variability— in particular due to a significant variability in UV radiation. Are there other important mechanisms to take into account as well?

JH: Well, earlier we mentioned cosmic rays. As we said, there are people who have measured cosmic rays at the earth's surface since the 1950s and it's clear that they are modulated by the solar activity—there are more when the sun's less active. And we know that cosmic rays are the main source of ionization in the upper atmosphere, so people have hypothesized that solar modulation of the ionization might have climate impacts, and there are two ways that this could happen.

One is through changes in the earth's electrical circuits: there is an electrical circuit in the atmosphere, so this might influence convection and thunderstorms and things like that.

The other effect that's been suggested involves the particles inherent in clouds. When clouds are created in the atmosphere, you need them to condense onto little particles—the water vapour won't just spontaneously condense out, even when it's below zero—you need tiny little particles of dirt or something floating up there for the clouds to condense onto.

And there's a suggestion that,when there are more cosmic rays and there's more ionization of these particulates in the atmosphere, so you might get more cloud—or at least, clouds made of smaller particles, which have different radiative properties—indeed, they're more reflective. So the basic idea is: more cosmic rays, more reflective clouds, cooler climate—that's another potential mechanism.

So far it has had much less validation or testing than some of the other more radiation-based theories.

HB: And presumably, there's an additional question once one flushes out these details a little bit more, which is, *"How do these different mechanisms precisely interact amongst each other?"* So, if you have two or three different mechanisms, you can imagine that they might sometimes compliment each other and sometimes interfere with one another, and so on—again sometimes in a linear way and sometimes in a non-linear way.

JH: Yes. One topic we haven't talked about yet is solar particles, but I did say that they were coming in at high latitudes—and they do: they cause the Aurora in the upper atmosphere, but when they get a bit further down into the stratosphere they noticeably cause ozone depletion, so you get ozone depletion events at high latitudes when the sun's more active.

And this depleted ozone air sinks downwards into the equator within the stratosphere. So, based on what I was saying before about changes in radiation affecting the ozone and affecting the climate, you can see that there's a plausible route whereby that coupling might also influence it.

They're much more intermittent than the 11-year cycle, and so the effects might last a little while, but it's not clear whether they come far enough down or whether they can have a significant enough effect on temperature to produce any effect at the surface—but in terms of nonlinearity you're absolutely spot-on there, because when the sun's more active, you get more storms depleting the ozone and more UV, which is increasing the ozone, and so it's a big mess.

Questions for Discussion:

1. Under what circumstances is a successful "forecast" more impressive than a successful "hindcast"?

2. Is there a certain type of scientific temperament that one needs to have in order to be a successful climate scientist? In what ways might this be different from other areas of scientific research?

3. To what extent might the "politicization" of climate change influence the research topics that a scientist might choose to study? What are the positive and negative aspects of this overlap?

VIII. Getting The Word Out

Increasing public awareness

HB: Let's move now to the man or woman on the street and their interpretation of what this all means.

The models may be complicated, but as you say they're the natural product of integrating a great deal of basic physics in a rigorous way, so that even though you often have to take into account instabilities and complexity and chaos and what have you, effectively what we have is a case of scientific rigour in action.

Sure, you need high-level models and advanced computer simulations but at the end of the day it's clear that real, objective progress is being made.

But on the other hand, I can imagine that it might at least sometimes be the case that the complexity and sophistication of these models involved is misportrayed, to the extent that the average person might just throw up his or her hands and say, *"Well, there are all these factors and possibilities in play so clearly nobody really knows what's going on and I'm just not going to worry about it."* Do you ever have to deal with that sort of a reaction?

JH: All the time, absolutely that.

I think one problem we have as scientists is that we frequently refer to uncertainty as something that is fundamental to the whole scientific process, but to a person who's not a trained scientist, uncertainty means, basically, that you don't know.

So if we talk about an uncertainty range in a forecast the layperson would just think that we don't know what we're talking about. I'm trying to stop talking about uncertainty now in my public communication of these issues and instead talk about risk and odds.

People understand that if you're, say, betting on a horse race, there are certain odds of something happening or not happening— they buy that. Whether the horse wins or not may not be physically determined, but there are chances of certain things happening and the chances of an old donkey winning are very low.

So we try to say things like, *By doing all this physics, we can look at the chances of the temperature being such and such and we have a 90% certainty*—well, I've used that word again—*that such and such will happen.*

HB: I'm guessing that this is an exceptionally difficult road for people in your position to navigate because if you speak to people as you would speak to your scientific colleagues and say, "*Here is our range of uncertainty and here is our conviction of these particular parameters or those particular parameters*", you're justifiably concerned that such a statement will be misinterpreted as, "*Well, these guys really don't know what's going on*"—which is not only profoundly untrue, but will also result in a dangerous level of inaction in terms of public policy.

But on the other hand, if you just wave your hands around and don't acknowledge the scientific uncertainties involved, then you run the risk of appearing decidedly unscientific and looking like a bit of a politician.

JH: Oh yes. The Intergovernmental Panel on Climate Change produces huge reports every so often on the state of the climate—what we know about it and what's going on. It's created by hundreds of scientists working together to write these reports, and they have to evaluate degrees of understanding or uncertainty ranges or whatever you'd like to call it during all their discussions.

And we—I haven't done it recently, but I've done it in the past— sit for hours discussing whether it's "probable" or "likely" or "not likely" or "extremely likely", discussing precisely which words to use and what those words actually mean in terms of what your models or your data are saying. I mean, people just talk for hours on it.

HB: I can imagine, but it's extremely important because if you have, through scientific rigour and analysis, come to the conclusion that you mentioned earlier—which is that there can be no doubt that the current changes in our climate are caused, not by these natural causes, due to solar variability or what have you, but by man-made, anthropocentric factors—then it's very important to clearly state that in the hopes that something can be done about it.

After all, we're not talking about whether or not we discovered this subatomic or that subatomic particle at CERN—we're talking about an issue that affects everyone and that there's a strong likelihood that we can directly influence, positively or negatively, in terms of some future public policy.

But on the other hand you don't want to misrepresent or trivialize the process of science itself by pretending that things are 100% black and white.

JH: Well, that's true, but that's inherent in science in general, I think. Most people would understand, for example, if you went to your doctor with some complaint and after you describe your symptoms, the doctor would say to you, "*I'm not entirely sure, but I'm pretty confident that it's such and such and I'd like to prescribe such a treatment for you*", you would understand what that meant. You would understand that the doctor couldn't be *completely* sure, but it was probably in the best interest of you to take this treatment, because, if you don't try to take it, something worse might happen.

HB: Well, perhaps I have more confidence in atmospheric scientists than I do in the medical profession—maybe that's just me and you're talking to the wrong guy. But as it happens, since you brought it up, *this* is precisely one of the problems I'm talking about, because in my experience many people in the medical profession *don't* actually admit when they don't know something and often seem more concerned about projecting an air of confidence and authority—and if, when I start questioning them, it becomes obvious to me that they don't actually know, they lose all credibility in my eyes because they've been bluffing.

Again, maybe this is just me, but I think what I'm trying to touch on is that, maybe one of the issues that you're dealing with is a lack of public understanding of legitimate open and honest scientific practice and what it means to be able to say that you don't know.

You talk about how people spend hours and hours around tables thinking about what words they should use to assuage the public and so forth but, from my perspective, I think it's an important part of the scientific practice to say, *"Look, we don't know **everything** and we don't **pretend** to because such posturing goes deeply against our principles and to some extent invalidates the claims we do make. But **this** is what we **do** know, and this is **why** we say that we know it; an, if that makes you feel uncomfortable, let me explain to you what science is all about."*

JH: Well, if anybody would stop and listen, of course, that's exactly what we'd like to say, or be able to say, but governments want you to give an answer because, obviously, they need to know what to do in their policy on, say, building nuclear reactors or whatever we're talking about this week.

HB: Well, I want to get to government and public policy in a moment, but let me talk first about one of my hobby-horses, which is the media, because we haven't talked specifically about that and it seems to me that the media plays a pivotal role here, because they are the primary avenue through which you communicate to the public.

If you're a member of the general public, how else do you find out? You're not going to go read a technical article on atmospheric science in some journal somewhere, you're going to watch a debate on TV, you're going to read it in the newspaper, or listen to an interview on the radio or what have you. Is the media doing its job and, if it's not doing its job as well as it could be, what should it be doing differently?

JH: Some aspects of the media are doing a really good job and they have really good environmental correspondents; they come and talk to us, they try to understand what's going on; and if they don't understand things, they'll consult the science media center and they'll ask several scientists about a particular issue to get the whole picture.

However, certain aspects of the media want an exciting story; and if somebody comes up with something that says, "*Oh, the sun is causing climate change*," they love that and they'll push it, just because it's more exciting than me saying, "*Actually, it's not doing very much*." And you can see why: they want to sell papers.

Certain journalists appear to have bees in their bonnets—politically or otherwise they seem to be motivated to support climate change denialist angles—and you'd have to ask them why they want to be like that.

HB: These are individual journalists, rather than certain types of media? Do you have a sense that you can meaningfully generalize between the different forms of media—television, broadsheets, tabloids, radio, social media, what have you?

JH: No, there are good and bad ones in all of the mediums, absolutely all of them. So, if you take the heavyweight newspapers, some are really solid and others are publishing this denialist and rubbish stuff—presumably because proprietors want them to say that sort of thing. I don't know why they would do that, but it does vary from journalist to journalist as well, and some of them are much better than others, even within each of those particular newspapers.

Then you get the tabloids, some of those publish pretty nice stories. Obviously, they're not in such depth or detail, but they get the essence and they put it over, so it's not so easy to distinguish between different forms.

In terms of radio: Radio 4 is perhaps the obvious one, but I've also done Radio 5 Live, Chatting Radio London—people seem to be interested, but perhaps they're just being polite. I don't know.

HB: Well, you are British after all—it can sometimes be hard to tell when politeness is the driving force.

But there *has* been a change in societal attitudes, right? There seems to be little doubt that there's a vastly increased amount of awareness for issues such as climate change. I'm sure there are very many deniers still, but I'm guessing that, relatively speaking, there are far fewer deniers now than there were 30 years ago or 20 years ago or perhaps, even 10 years ago.

So, something seems to be happening in terms of public awareness. Maybe it's not an optimally efficient system of communication, but the scientific establishment's word *is* getting out; people know about the IPCC's reports, and there is a general understanding that a growing and mounting body of scientific evidence is pointing in this particular direction, wouldn't you say?

JH: Yes, I think that's right, and there have been some really good programs in schools as well. Some people might say that the children are being brainwashed, but I believe that the environmental science that's being taught in schools is so much better than it ever was before—the children are educating their parents, to a certain extent, so that's very good.

Questions for Discussion:

1. *Do you agree with Joanna that re-expressing scientific predictions in terms of "odds" rather than "certainty measures" is a better way of communicating the ideas to the general public?*

2. *Do you think that, on the whole, the media does a responsible job at communicating scientific ideas to the general public? How could it be improved? Is the media's coverage of climate change treated differently than other scientific issues?*

IX. Public Policy

From words to acts

HB: Moving on to the political arena, you pointed out that politicians have to know what to do at some point. As a politician, you can't just say, *"It's all quite complicated, but I have a general conviction that this is the prevailing consensus on this issue"*. You need to regularly make concrete decisions like where, precisely, to site a nuclear reactor or whether one should be sited at all.

So here's a direct question, finally: If you were prime minister with a majority government in this country, what would you do? What sorts of policies would you invoke?

JH: I would put much more investment into renewable energy. At the moment, that sort of energy is not as cheap as oil or coal, but as the technology improves that will change—we've seen the price of solar photovoltaics, for example, plummet.

There's such an opportunity for innovators and people to invest in new technologies that will create clean energy in any number of clever ways that I can't think of but these inventors can—some of them are already in progress and others are waiting to be discovered. Invest strongly in that, is what I say

Meanwhile, we can solve half the problem by just making our energy use more efficient. So, just doing that is what they call low-hanging fruit; it's an easy thing to do, subsidize that sort of thing.

HB: OK, so that's Joanna as Prime Minister. What about if you're not in a position of enormous power and responsibility? I can imagine someone saying, *"Well, this is all very interesting, but what can I do? I'm just one person somewhere. I'm not actually prime minister, I'm*

not a major investor in any companies, fossil fuel or otherwise. Is there anything that I could be doing, not only in terms of my pattern of consumption and my day-to-day living, but also in terms of education, awareness, political lobby or what have you?" What advice might you give to the average person?

JH: I think it's very important to lobby your local political representative. Before the last election, I talked to lots of political candidates and they told me that if many of their constituents were telling them that they were worried about climate change, they would be much more likely to do something about it than if it was just these old scientists banging on about it.

That's the number one thing to do, I think: get all your friends to tell your local representative that you're concerned and ask if they can do something. And try to make sure that your local council's doing good stuff with recycling and energy saving and local transport, improving local transport—it's all the obvious stuff.

HB: Are you optimistic about the future?

JH: If you look at the international perspective, it's really encouraging. What happened at the United Nations climate meeting in Paris was amazing, it was so extraordinary. You had 195 countries unanimously declaring that they were going to do something about it. If you think about how different that was from Copenhagen six years previously, where they couldn't agree to anything at all, it's just amazing. I wouldn't have known they were going to agree on anything, let alone that, so that's hugely positive.

Subsequently, there are countries that have all signed up to it and it looks like they're really going to try; that's really positive. Then, you come back down to earth and you think, *"My goodness, what and how are they going to do anything?"* and you get the British government, which has done nothing—it's taken away subsidies for renewable energy; it's stopped the Carbon Sequestration Program; it's stopped the Green Deal; all it's doing is going along with a Chinese and French nuclear reactor, and that's not because they're interested in climate change—they use that to say that they are—but they're really only interested in that for an easy energy supply.

So, from the UK perspective, it's not good. However, one caveat on that statement is what the UK does have, which is really good, is the Committee on Climate Change. This is a panel of independent experts that provides for plans for what the carbon emissions of the country should be in five-year chunks into the future.

The carbon budget is now set for up to 2030, going down to, I think it's 57% of what it was in 1990 and they have been passed through laws, so they have gone through parliament and it's actually legally binding on the government to do it.

The other big issue is energy costs: if people have to pay more for their energy then they're less likely to want to change, so if the government can subsidize renewables until they achieve parity with fossil fuels—which they will do—then that would be a very good thing.

Questions for Discussion:

1. Which country do you think has the most progressive and impactful policies to encourage a rapid transition to green energy?

2. How do you think public policy on the environment would change if ⅔ of the politicians in the country had a scientific background?

3. Should climate change be regarded as a "political issue"?

X. Final Thoughts

Towards a better future

HB: I have two final questions, you've been very generous with your time, thank you very much. To begin, I'm going to tack back to the science and ask a favourite question, which is, *If I were an all-knowing being and I could answer any scientific question that you would have, what would you ask me?*

JH: Well, there are so many little things aren't there? However, I suppose you'd want to ask a big thing.

HB: It's entirely up to you—in fact, you can ask more than one question; I'm very generous today. You can even be particularly sneaky and ask for an infinite number of questions. I'm just curious to know what sorts of things you're dying to know.

JH: Well, I would like to know why people want to be climate change deniers—what's in their psychology? I mean, obviously some people are being paid by the coal lobby and that's obvious when they've got land with coal or oil on it, but there's a certain brand of person—including scientists, actually, they tend to be retired geologists—who want, somehow, to deny the science and I'd like to know why. Please, can you tell me?

HB: Well, that's not part of the deal, unfortunately—I can't actually do anything, I just wanted to know your response. But as it happens that was actually something I was meaning to ask you and had forgotten, because that had perplexed me too: there are people with personal financial interests and then, there are people who might be sufficiently sceptical based upon their particular experiences. But there is this large middle ground of people who clearly don't have any obvious economic interests and who don't have sufficient knowledge of the particulars of the field and yet, those people tend to be very vehement in their denial.

JH: Yes, and there is a section of the press that feeds and encourages that and talks about climate scientists as sitting on fat grants and getting paid to tell everybody lies so that we can just enjoy our green, lefty agenda—whatever that is.

I get loads of letters and emails asking questions, some of them say, "*You've got it all wrong*" about something or another, and I always write back once and try and say, "*Look, this is actually how it is*," and usually, I don't get anything back.

Some of them do get back to me and go on and on. And about one in twenty will write, "*Thank you very much for explaining that; I didn't understand it before*," and I think, "**Yes!**"

HB: That's not bad; one in twenty.

JH: Well, I don't know what the real success rate is—maybe it's not that high, actually, but anyway it's worth trying. Other people are just abusive, but that's their problem, not mine.

HB: My last question concerns education. It might be related to climate science and it might not be, but what advice would you give to teachers and educators with respect to science education more broadly?

You mentioned earlier the formative influence a high school physics teacher had on you, so clearly you appreciate the importance of this. Are we doing a good enough job, in general, educating our children with respect to science? How can we do a better job? What sorts of things might we be encouraging or discouraging?

JH: I think it would be helpful to try to convey more understanding of what's going on in the world. You can do that from many different aspects; you can do it from actually understanding the science—physics, chemistry and biology—and also on the impacts by looking at the outside world and saying, *"Look, you're seeing this happening here"*.

I think they actually do that better now than they did when I was at school, where it was definitely more about equations. Personally, I quite like equations, but the point I'm trying to make is that demonstrating to children the real-world aspect of science is important, and—going back to the question about why girls don't generally do physics—it might help that agenda too, because I think girls like to feel that they can actually make a positive contribution to the world.

HB: Would you have any specific advice towards a young person who is considering a career in climate science and environmental science? Would you have any words of wisdom that you would like to bestow upon them?

JH: I don't like giving advice, but I would say to them, *"If people offer you advice, then listen to it and weigh it up, but recognize that other people have different reasons for telling you things; I know, from my own experience, that if you just believe everything you're told, you might go down a wrong track. So, by all means, take all the advice, listen to it, but don't necessarily go by it; do what you feel your gut instinct says is right for you and that will probably be the better way to make decisions."*

HB: Sounds good. Anything I haven't asked? Anything you'd like to add or we should get back to or emphasize?

JH: I don't think so, I mean, I could talk forever about atmospheric science and all these poor people sitting around the set will fall off their chairs in boredom.

HB: I don't think so. I mean, the accomplishments in the field are just amazing. One of the things that impressed itself upon me was just the sheer amount of knowledge of past events that we have these days. I mean, getting a handle on the climate of 5,000–10,000 years ago? That's one of these things that I would have thought would be impossible.

JH: Yes, so in our discussion we talked a lot about models but much less so about observations—and, of course, they're hugely important: you can't have models sitting in a vacuum not relating to what's actually happening. It's very important to appreciate the observational record going back in history and all the data that we're getting from satellites now on what the atmosphere is doing all the time—not to mention the oceans. We haven't talked about oceans, but they're very important too.

HB: Well, thank you once again, Joanna, for all your time and your detailed explanations. It was a real pleasure.

JH: You're welcome—it was a pleasure for me as well.

Questions for Discussion:

1. Do you agree with Joanna that by deliberately stressing the potential real-world impact of the science more girls will be attracted to studying physics? Might it also increase the number of boys studying physics?

2. In what ways has this conversation given you a deeper appreciation of the personal and professional challenges of a climate scientist?

Continuing the Conversation

Those interested in a more detailed understanding of solar impact on climate are referred to Joanna's book co-authored with Peter Cargill, *The Sun's Influence on Climate*.

Ocean Enlightenment

A conversation with Edie Widder

Introduction

From Sea to Shining Sea

The first reaction anyone has when witnessing the spectacular light show of marine bioluminescence is that it is overwhelmingly beautiful. Scientists, of course, are typically expected to be much more clinical and detached in their personal reactions to natural phenomena, limiting themselves to a noble satisfaction at unlocking the mysteries of the world around them—the pull of the truth, the thrill of discovery—that sort of thing.

But Edie Widder isn't your average scientist, and she feels no compunction whatsoever to disguise her very first thoughts when she found herself deep under water in a clunky deep-sea diving suit surrounded on all sides by beautifully glowing objects:

"All I could think was, 'Oh, wow! This is just so incredibly cool!'"

Widder, naturally enough though, didn't stop there. Equipped with a PhD in neurobiology and a steely sense of logical rigour, she swiftly pushed her natural exuberance in a more concrete direction.

*"But I also recognized just how much energy was involved in producing all that light. And I thought to myself, **This is a crazy amount of energy. This has got to be one of the most important processes in the ocean—why aren't more people studying it?**"*

So began Edie Widder's love affair with bioluminescence—a passion that led her to become a leading specialist in the field while simultaneously developing a broad spectrum of pioneering devices: an instrument for the open ocean that the US Navy could use to minimize its bioluminescent footprint or locate that of others, an ultrasensitive deep-sea light meter for use in the deep ocean, and a remotely operated deep-sea camera system known as the *Eye-in-the-Sea*, which she used to record an entirely new species of squid by using a specially designed bioluminescence lure.

But bioluminescence isn't just brilliantly useful, it is also a fascinating scientific phenomenon in its own right that naturally steers us towards a much deeper understanding of fundamental biological processes.

Of course, emitting light costs an organism a good deal of energy. For larger objects like shrimp or jellyfish, the extra energy required by bioluminescence is clearly justified by an enhanced ability to find food, attract a mate or defend itself from a predator. But when we consider much smaller bioluminescent organisms like bacteria, the reason for spending so much energy to create all this light becomes much harder to justify. In fact, if you combine groups of bioluminescent bacteria with their non-bioluminescent cousins, you will quickly find that the non-bioluminescent bacteria will emerge triumphant due to the extra energy that they save by not producing light. So if bioluminescent bacteria can't seem to battle their way past their darker compatriots, how on earth could they ever have gained the evolutionary upper hand?

The answer to this long-troubling conundrum, Edie told me excitedly, only emerged relatively recently when careful studies by Polish scientists revealed that only bioluminescent bacteria have what it takes to repair potentially damaged DNA.

"So the original adaptation for light output in the bacteria was DNA repair—suddenly it all makes sense. It's beautiful."

Stimulating stuff, all the more so when coupled with Widder's overwhelming capability to combine being a top-flight researcher, expert deep-sea diver, experienced submersible pilot, master engineer/technician and bourgeoning media star.

But however fascinating and compellingly beautiful I found biolu-minescence, and however personally impressive I found Edie, there's much more than that going on at Ocean Research and Conservation Association (ORCA), the not-for-profit that Edie co-founded back in 2005 on the back of her MacArthur "genius" Fellowship.

In South Florida's Indian River Lagoon, under the watchful eye of Widder and her fellow ORCA staff, local high school students were engaged in vital scientific research to monitor coastline water quality, developing precise pollution maps of their environment by rigorously testing soil sediments with biologically luminescent bacteria.

Why high school students?

> *"As I become more and more of a conservationist and begin to take the 30,000 foot view of things, I realize that the only thing that is more unconscionable than destroying these ecosystems as we have is handing ecosystems that are spiralling out of control to the next generation without giving them the tools to deal with the situation.*

> *"When I was born there were 2.5 billion people on the planet. There are 7 billion now. That's just mind-boggling. All these people are using resources and creating waste and burning fossil fuels and we're creating a chemical Armageddon as a consequence.*

> *I mean, the fact that we could actually be **acidifying** our oceans is simply mind-boggling. And the toxins that are pouring into our water-ways and poisoning ourselves as well as all of these ecosystems—how are these kids supposed to deal with this?"*

How, indeed?

Somewhat surprisingly, optimism is the key. Less surprisingly perhaps, science must lie at its heart:

> *"The first thing we have to do with these kids is actually teach them optimism. And the way you do that is to do it realistically, because blind optimism in and of itself is not a solution.*

*"They have to know that they are problem solvers. I think this all harkens back to the way we teach science in our schools—which is very, very badly. I mean, we actually teach science by having kids watch a candle burn and write down their observations. Now I know a really gifted teacher **might** be able to make that work, but honestly that's about as exciting as watching paint dry.*

*"We've got kids at an age when they want to make a difference in the world. They're passionate. Let's harness that for **real** change in the world. Let's get them involved in doing these measurements that nobody else is doing, solving problems that nobody else is solving, learning how to solve problems."*

The bad news is that, once the measurements are actually made, the problems may be even worse than we had feared. In the case of the Indian River Lagoon, recently constructed pollution maps show a mysteriously high level of toxicity that is far beyond what anyone had imagined.

But optimism still reigns. Edie is now turning her boundless energy to making her program sufficiently scalable on a regional, national and international level, trying to find a simpler, lower-cost pollution test that can be used by those who don't have access to expensive lab equipment or advanced scientific expertise.

"My ultimate goal is to actually make these available on the web as open source material and then have other people feed into it. There are citizens' science groups where kids take on civic responsibility issues. Other groups could come in to this from all different angles."

Imagine a world where the next generation could successfully work together to clean up the mess we've made of our oceans and coastlines.

Now *that* would be incredibly cool.

The Conversation

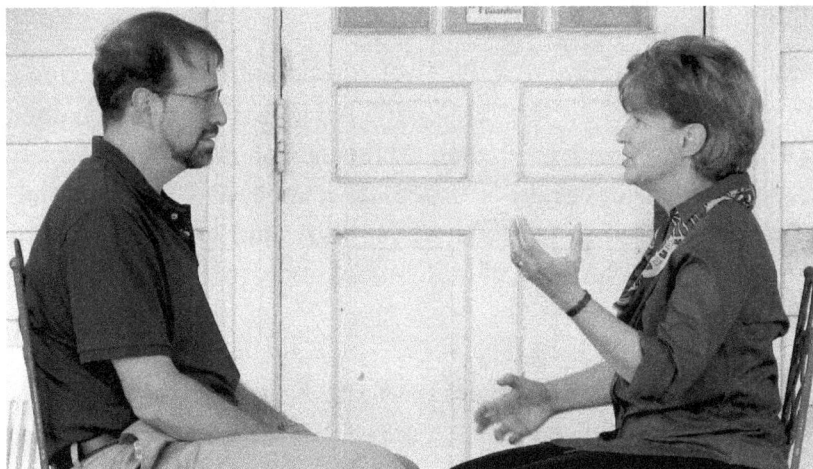

I. Bright Beginnings

Underwater awesomeness and environmental opportunities

HB: There's a lot to talk about here, but let me begin on the basic science side of things. How did you first become interested in bioluminescence?

EW: Well, when I started my PhD work, I was offered this opportunity to work on a bioluminescent dinoflagellate with a woman named Beazy Sweeney, who was famous in the field of phycology, the study of algae. My major professor was Jim Case, but he and Beazy Sweeney had gotten together and thought up this project of making electrophysiological recordings of bioluminescence from a dinoflagellate that she had in culture.

I remember having this long conversation with Beazy Sweeney about bioluminescence, where I was sitting there trying to nod and look intelligent while thinking to myself that I don't really know what bioluminescence actually is.

I rushed home that night and pulled the Encyclopedia Americana off the shelf—because this was the old days—and looked up bioluminescence, and it turned out that the entry was written by Beatrice M. Sweeney.

HB: So you were in the right place.

EW: I was, although I knew nothing when I started. Jim Case was quite famous in bioluminescence research with fireflies. So yes, I was in good company. As it happens, bioluminescence itself was not actually the focus of the research. It was a convenient effector system to record from—I only gradually got more and more interested in this ability of creatures to make light.

Jim was brilliant at writing proposals, and he wrote a successful proposal to the American Department of Defense for an optical multi-channel analyzer—which is a fancy way of saying spectrometer—that was so super-sensitive it could measure bioluminescence spectra, the colour of the light.

He got the equipment and it was just sitting there on the bench in the lab. I've always been kind of a gadget freak, so I couldn't quite leave it alone. I just kept fiddling with it until I pretty well became the lab expert on it. Then one day he turned to me and said: *"Well, now that you know how to make this thing work, we need to start sending you to sea to measure all these animals in the ocean that make light that nobody has ever been able to measure before."* Suddenly I was a seagoing marine biologist—which is what I'd always wanted to be—but never thought anybody actually **gets** to be; and I loved it.

But even at that stage I didn't really think you could make a career out of that. I actually had a postdoc lined up in Madison, Wisconsin, with the leading membrane biophysicist of the day. It was a dream postdoc for me. And then I went out on this expedition where there were some scientists testing this atmospheric diving suit called the WASP. It was developed by the offshore oil industry for diving off oil rigs down to depths of 2,000 feet, and they were testing it as a tool for exploration.

HB: What does it look like?

EW: It's got Michelin-man arms but no legs, and kind of a pod and thrusters that fly it around. It's called the WASP because somebody thought it looked like the insect: a yellow body and a big bulbous head.

During this first expedition, I'd be on a headset talking to whoever was in the suit. And I'd say to them: "*Would you turn off the lights and tell me what you see.*" Of course I knew they'd see bioluminescence, which had been known about for a very long time.

So they'd turn off the lights, and all I'd hear would be comments like, "*Oh, wow! That's awesome!*" and I would ask them, "*Could you please be a little more specific?*"

And they really couldn't, which was very, very frustrating. So the chief scientist on the mission, Bruce Robison, took pity on me and said: "*Well, if you want to stay around for another year, we could probably get you trained up as a pilot, and you could get down and see for yourself.*"

On the basis of that promise, I turned down the dream postdoc in Wisconsin and lifted weights for a year to be able to use these incredibly cumbersome Michelin-man arms on this suit.

Finally, I made my first dive in the Santa Barbara Channel one evening. I went down to a depth of 880 feet, turned off the lights, and all I could think was, *Oh wow! This is just so incredibly cool!*

But I knew enough at that point to recognize just how much energy was involved in producing all that light; and I thought to myself, *This is a crazy amount of energy. This has got to be one of the most important processes in the ocean, why aren't more people studying it?*

I've been utterly hooked ever since. Fortunately for me I was actually very well-funded because the US Office of Naval Research had a very strong interest in bioluminescence due to its ability to reveal large cigar-shaped objects moving through the ocean at night.

HB: I see. So there was a military use.

EW: Yes, there were military applications. So I ended up being involved with Jim Case in developing what is now the Navy standard for measuring bioluminescence in the world's oceans: I co-hold the patent on it.

Believe it or not, a lot of the development of that instrument took advantage of my doctoral work studying the electrophysiology of that dinoflagellate since I understood the excitation mechanisms and what it was going to take to be able to control the excitation in an instrument that could actually give us a meaningful measure of the light output.

What the Navy wanted was to have a predictive capability for bioluminescence. They wanted to have an instrument that they could take around and be able to say, "*OK, you are working in this part of the ocean at this time of the year, so you have such and such a probability of there being bioluminescence there at this time.*"

HB: Which would presumably guide you towards certain efforts to minimize the amount of induced bioluminescence so that other people couldn't find out what you were doing.

EW: Or if they were looking for Soviet submarines, then they'd be looking for bright spots. But when the Soviet threat went away, my funding went away with a pop.

But then it suddenly came back during Desert Storm, as one of the SEAL teams actually had to change its insertion point because of bioluminescence, in order not to light up like Roman candles under darkness. So it all spun up again, but now it had shifted from an interest in bioluminescence in the open ocean to bioluminescence in the coastal environment.

Oddly enough, that's kind of where I got into the conservation aspect of things, because it coincided with the reports that came out in 2003 and 2004—the Pew Oceans Commission Report and the US Commission on Ocean Policy Report—which talked about the really desperate state of our oceans.

It was a scientific consensus unlike any I had ever seen before, and a big emphasis in both those reports was the need for better monitoring.

And when I was looking at the Navy's problem of how to develop a predictive model for the coastal zone environment, I did a little back of the envelope calculation to figure out how much money it would take to measure bioluminescence and develop that predictive capability right here in the Indian River Lagoon. I was just floored at the cost—I mean, it was astronomical even by Navy standards.

And it struck me that it was crazy that they wouldn't have any way to monitor the most basic aspects of our environmental health. So I thought to myself, *Well, maybe that's something I can actually help with, because I've spent a lot of my career working with engineers to develop solutions to questions that I wanted to answer in the deep ocean, so maybe I could do the same thing in the coastal zone environment and address some of these issues and develop new technologies that would make it easier and less expensive to do that kind of monitoring.*

Questions for Discussion:

1. Are you surprised to learn that Edie didn't think that she could make a career out of being a seagoing marine biologist? Why do you think that is?

2. Do you think that the state of our oceans has a lower public profile than other environmental issues?

3. What are the advantages and disadvantages of basic science being funded by the military?

II. Bioluminescence

Evolution in action

HB: I very much want to get back to that, but first I want to explore a bit more about the science of bioluminescence. You threw a lot of words around at the beginning—like "dinoflagellate". What's that, exactly?

EW: Dinoflagellates are algae, they're plankton. If you've ever gone out on the ocean at night and turned off the lights, you might have seen luminescence in your wake. Unfortunately where most people see it for the first time is actually in the "head"—which is the name for the toilet for landlubbers—because toilets on ships are flushed with unfiltered seawater that often has bioluminescent plankton in it. And if you find yourself staggering into the head late at night and you are so toilet-hugging sick that you don't turn on a light, that can be very disturbing.

At any rate, dinoflagellates are a very common source of bioluminescence, especially in the coastal zone environment. But as you move offshore, there are more and more animals that make light. If you went and dragged the net out in the open ocean environment virtually anywhere in the world from 1,000 meters to the surface, most of the animals you'd bring up in that net would make light: most of the fish, the squid, the shrimp and the jellyfish. In some places 80% of the animals make light.

HB: That's just fascinating. I was vaguely aware of this—I knew about some plankton and so forth—but I had no idea that so much of marine life in the open ocean is bioluminescent. So why is that? And what is bioluminescence anyway?

EW: Well, for most people, if they know about bioluminescence at all, it's through fireflies. There are a few other land animals that make light, but in general it is pretty rare on land. I think that's why people think the same thing is true in the ocean as well. It isn't, as it happens—but that's the thinking.

As a physicist, you can appreciate that all light in the universe comes from the same basic phenomenon: an electron gets excited to a higher energy level, and then drops back down to a lower energy level while giving up its energy as a packet of energy called a photon. The only difference between different types of light is how the electrons get excited in the first place.

Thermal excitations are called incandescence, and candle flames, sunlight, and many other lights we see—like the ones you are using to film right now—all produce both heat and light. But there are other ways you can get electrons excited, and one of them is with a chemical reaction, in particular a very efficient chemical reaction that's called chemiluminescence. Bioluminescence is just one form of chemiluminescence, where the animals are producing the chemicals that make the light.

HB: So this is an internal process caused by some chemical reaction. And from an evolutionary perspective, there's presumably some justification for why so many creatures are doing this, and doing it in such a large variety of ways. Tell me a little bit about that.

EW: The three broad categories affiliated with bioluminescence are finding food, attracting mates, and defending against predators. As for why there is so much bioluminescence in the ocean, our best guess is that as the ocean filled up with ever swifter and nastier predators, prey needed to be able to escape those predators, typically by hiding. But there are no hiding places in the open ocean: the only thing to do is to go down deeper where it's darker.

These are animals that had already evolved eyes, and now the evolutionary pressure is to develop more sensitive eyes, or more sensitive visually-signalling capabilities, because they're living down deeper.

Imagine an animal that already had developed, say, spots on the opercula for a fish to attract a mate, and now it's living deeper to escape its predators and it's got to make those spots more visible.

So that's the evolutionary pressure for it, that's why you get animals that have developed built-in flashlights to help them see in the dark, glowing lures to help attract food to them.

Some of them are chin barbels; sometimes it's arched in front. We also find different colours of light being generated, which is very interesting. Most bioluminescence is blue because that's the colour that travels furthest through seawater, so animals have evolved the wavelength that's going to travel the furthest, and most of them can only see blue light as a consequence.

But, for example, there is a deep-sea fish that I love that produces far-red light from a big red light organ under each eye. It has visual pigments that allow it to see red light. It uses this red light like a sniper scope, so it can sneak up on animals that are blind to the red light.

HB: These are for shorter distances, presumably?

EW: Yes, right.

HB: So it's got this extra visual capacity that other animals don't have. It's like a Navy SEAL with night-goggle vision.

EW: Exactly. The colours range all the way from ultraviolet to far red. Most of it, as I said, is blue. As you go into the coastal zone, it tends to be greener because the light is scattered more by particles in the water, so green light actually travels further and they've evolved *that* colour to predominate in the coastal zone environment.

HB: That brings up something else I was going to ask: you've talked about your work in bioluminescence in the open ocean and in coastal areas, and I wanted to ask you if there's any difference between them.

EW: There are differences in colour, but there is a huge difference in the amount of species that are bioluminescent. As you go offshore, as many as 80% of the species that you bring up could be bioluminescent. But in the coastal zone environment it's generally only about 10% of the species that are bioluminescent. And that's because in this local environment there are plenty of hiding places and the evolutionary pressure to develop bioluminescent properties wasn't nearly as strong.

Bioluminescence gets used a lot for defence in a lot of different ways. Just like the way a squid or an octopus will release an ink cloud in the face of a predator, there are many animals that can release their bioluminescent chemicals into the face of the predator temporarily blinding it while they swim away into the darkness.

Others use it for camouflage. Hiding in that dim, downwelling light they've got to worry about their silhouette, so they produce luminescence from their bellies that exactly matches the colour and the intensity of the downwelling sunlight, and it is so perfect that if the cloud goes over the sun and dims the sunlight, they dim their belly lights and they truly disappear.

HB: They can adjust in real time to variations of sunlight?

EW: They've got a feedback system.

HB: That's amazing.

EW: Sometimes there are individual spots on the belly, but if you've ever opened your eyes underwater you know how things blur, and it blurs together perfectly.

Fish do this, some sharks do it, shrimp and krill—squid do it spectacularly. In fact, there are some squid that can change the colour of their luminescence depending on whether they are trying to camouflage themselves against sunlight down deep during the day or against moonlight up near the surface at night.

HB: No!

EW: Yes, and that's a *really* clever trick as it's under temperature control: if they're down deep during the day it's cold and when they come up in the surface waters at night it's warm. You can actually make them do this in the lab by just changing the temperature and they turn on different light organs—in fact, it's even more complicated than that because some of these light organs have interference filters in front of them to change the colour a little bit and they can change the spacing on the filter in order to broaden the spectrum.

HB: So for anyone who doesn't believe in evolution—well, you either have to accept that there are these remarkably sophisticated, intelligent marine life forms out there that are somehow nothing less than Einsteinian in their adaptive capacities... or recognize that you're wrong.

EW: Bioluminescence is actually a marvellous way to teach people about evolution. For example, we discovered a deep-sea octopus that ended up on the cover of Nature because it was evolution caught in the act—it had suckers that were no longer suckers—they were suckers turning into light organs. You could see the light organs but you could still see the vestigial muscles, although they didn't exist as suckers anymore.

Another very nice relatively recent example concerns the discovery of an evolutionary mechanism for bioluminescent bacteria. When I was starting out in this field, every bioluminescence group that ever got together would end up in an argument about how bioluminescent bacteria could have ever evolved in the first place, because the light output from a single bacterium producing light can't be seen by any known eye. So how could it have been selected for in the first place?

In fact, the light output is so energetically costly that it seems detrimental to the bacteria to produce this. You can demonstrate that by taking a culture of bioluminescent bacteria and a dark strain of the same species and putting them together—the dark strain will overrun the culture because it has more energy than the bioluminescent bacteria that spend so much of theirs producing light. So it seems pretty mysterious how it could have got started to begin with.

But there were these very clever Polish scientists who did an experiment where they irradiated the dish of these combined bacteria with ultraviolet light, and then the *bioluminescent bacteria* overran the culture.

The explanation is that there is an enzyme called photolyase that repairs broken DNA, but it requires light of exactly the wavelength that the bioluminescent bacteria produce. So now we believe that the original adaptation for light output in these bacteria was DNA repair. And suddenly it all makes sense. It's beautiful.

HB: That brings me to the matter of open scientific questions in the field of bioluminescence. Are there any deep mysteries about bioluminescence that you'd dearly love to know? If you could talk to some all-knowing creature and have the opportunity to discover the answer to anything about this phenomenon, which specific questions would you ask?

EW: There are so many of those that it's hard to know where to begin, because an awful lot of what we think we know about bioluminescence is guess work and there's been very little direct observation.

So, for example, why does bioluminescence sometimes come in the colours that it does? There are even some examples that extend down into the ultraviolet region. And then there are some shrimp that we've discovered—working with a colleague of mine who measures the spectral sensitivity of the eyes of these animals—that can see ultraviolet light but their own bioluminescence isn't ultraviolet as far as we know. We've actually worked rather hard to try to figure out why this shrimp can see both blue light and ultraviolet light: what it's using that for. So far that's a total mystery that we haven't figured out.

There are many other little mysteries as well. There's this fish called the shining tubeshoulder that squirts out light just the way other animals squirt it out to defend themselves, but for some reason this fish doesn't just squirt out the luminescent chemicals, it squirts out whole cells, cells with nuclei and membranes. That seems energetically *insane*. So, **why** on earth does it do that? What is it that requires it to do that that's so different from what other animals do?

There are endless questions like that.

Questions for Discussion:

1. Do you think that most people would be surprised at learning that roughly 80% of creatures who live in the open ocean are bioluminescent? What does this indicate about the current state of how we teach biology if such a common phenomenon is so underappreciated?

2. If the original adaptation for bioluminescence was linked to DNA repair, what might that imply about the frequency of harmful mutations to these bacteria in that environment?

3. What does Edie mean, exactly, when she describes the shining tubeshoulder's behaviour as "energetically insane"?

III. The Eye-in-the-Sea

Glimpsing the unseen

HB: You mentioned earlier when describing the way you became involved in bioluminescence how you had a strong interest in gadgets and building machinery. During one of your TED talks you speak at some length about how you built the *Eye-in-the-Sea* camera to study life in the ocean using bioluminescence. I found that quite amazing for two reasons.

First, because I never would have thought that bioluminescence could be used so directly to study the sea. And secondly, because I had no idea that we know so little about how many species in the oceans actually exist. I had naively imagined that we have a very good idea of the extent of creatures that live in the sea, but it turns out that that's not really the case.

EW: The statistic often bandied about is that we have only explored about 5% of the ocean, and the sound bite that you hear most often is that we have better maps of the dark side of the moon than we have of the bottom of the ocean.

So there's that, but then there's the *way* we've explored the ocean. The primary way we know what lives in the ocean is either by dragging nets behind ships—and I defy you to name any other branch of science that still depends on hundred year old technology—or we go down with remotely operated vehicles or these very bright noisy platforms called submersibles. I've made hundreds of dives in submersibles myself, and as I've sat there with these lights shining out into the darkness, I've always wondered what was out there just beyond the range of my lights that could see me but that I couldn't see. I mean, any animal with any sense is just going to get the heck out of there. So I've had this feeling for a long time that I wanted to explore the ocean unobtrusively.

I actually got this idea from that deep-sea fish I mentioned a few minutes ago that uses red light. It can be done—it's harder, but it *can* be done. You can't use infrared light the way they do on land because it just gets absorbed too quickly. But you can use far-red light and a supersensitive camera. I wanted to be able to bring in more than just scavengers, attracting active predators by using bioluminescence.

So we developed what we called an electronic jellyfish that imitates certain bioluminescent displays that I had reason to believe would be attractive to large predators.

One of the ways that animals use bioluminescence for defense is something called the "bioluminescence burglar alarm". If you are caught in the clutches of a predator and have no apparent hope for escape, one thing you can do is attract a lot of attention in the hopes of attracting something larger that will attack whatever it is that is attacking you, giving you an opportunity to escape. This is actually a pretty common trick—it's the same reason that birds and monkeys have fear screams. They scream once they're caught by something in the hopes of attracting something bigger that will attack whatever it is that is attacking them and allow them to get away.

Animals will use bioluminescence for a lot of different reasons, but they will use every light organ they've got to try to attract attention if they're caught. There's a common deep-sea jellyfish called an Atolla—a beautiful thing—and it produces the most spectacular burglar alarm where the light just pinwheels around the jellyfish for an extended period of time. It is very, very bright.

So we created a copy of that bioluminescent burglar alarm by building a device with 16 blue lights that can be programmed to imitate exactly those spiral paths.

When I originally did this, I was trying to get it funded. And there's this problem in science that they won't give you money for something unless you tell them what you're going to discover ahead of time.

So I couldn't get it funded—and what I eventually did was to go to the Engineering Clinic at Harvey Mudd College, where I convinced them to do it as an undergraduate student project. They put something together that sort of worked on the bench, and then I went to the National Oceanographic and Atmospheric Administration (NOAA) and got enough funds to put the whole thing in a deep-sea housing and create a frame with a light. We cobbled it together by casting it in a mold of epoxy—you could actually see the word 'Ziplock' at the top of the electronic jellyfish because of the mold that we used to cast it in.

Then I got the Monterey Bay Aquarium Research Institute, where I'm an Adjunct Professor, to pay for a deep-sea battery. The first test was on a NOAA-funded expedition to the Gulf of Mexico. We put it down at a magnificent place called the Brine Pool, which is an underwater lake at a depth of about 1,800 feet. We set our underwater *Eye-in-the-Sea* camera down next to the edge of the Brine Pool, because it was like an oasis on the bottom of the ocean where I thought there might be a lot of large predators patrolling. We had the electronic jellyfish out in front and we left it down there for overnight recording. For the first four hours it was just recording the animals swimming around being natural with no stimuli.

And four hours in, I had programmed the electronic jellyfish to come on for the very first time with its pinwheel display. And **86 seconds** after we turned it on for the first time we recorded a squid over 6 feet long that's so new to science that it couldn't even be placed in any known scientific family.

HB: In just 86 *seconds*?

EW: 86 seconds. As you know, that almost never happens in science. But I certainly couldn't have asked for a better proof of concept, so I went back to the National Science Foundation and told them, "***This is what we will discover.***" And they gave me half a million dollars to do it right.

HB: Imagine if you'd found something after 36 seconds. They might have given you 2 million dollars.

EW: Right.

HB: So just to recap: you've got this electronic, simulated jellyfish emitting these light signals that effectively say, "*Help, I'm under attack!*" and *86 seconds* after you turn it on, out comes this big squid that no one has ever seen before that comes looking to eat the thing that it thinks is attacking the other thing.

EW: Yes.

HB: And of course it doesn't find anything because there's nothing actually there. You probably have a pretty frustrated big squid on your hands.

EW: Actually, we have examples of exactly that once we started doing it right. We put the world's first deep-sea webcam into the Monterey Canyon for about eight months and performed many sets of really proper electronic jellyfish experiments. There were lots of squid attacks on this electronic jellyfish and there were a number of times when it came in and—well, I know it's anthropomorphizing, but you could just see the frustration of the squid as it would come in and then go "*Whoa!*" then stop and back off and try again and then back off once more. Then it would try again from a different angle. On and on this went over and over again. So yes: they were clearly looking for something else than what they saw.

HB: It was obviously a huge challenge to get things going at first, but thanks to your initial 86 seconds discovery, you were quickly able to build a better quality *Eye-in-the-Sea* to concretely point the way towards what sorts of things are actually out there.

EW: It's just more proof that there's so much left to be discovered in the ocean that we've just barely begun to scratch beneath the surface. We have to be looking at new and innovative ways to explore if we're really going to find out what's out there.

HB: Do you have a sense that more and more people are working on these issues?

EW: No, actually. Less and less.

HB: Less and less?

EW: Yes, because of a lack of funding. Marine science has been traditionally underfunded and it's gotten much worse recently. It's very, very hard to get cruise time, ship time or submersible time. The one bright spot on the horizon is that now there are companies developing submersibles for the "entertainment market", as it were—people who want to buy a submersible for their private yacht. You can actually buy a very nice submersible for anywhere from $1–3 million these days.

HB: Quite a bargain.

EW: It *is* a bargain, actually, compared to what a research submersible used to cost. And these are really pretty impressive platforms.

So maybe through that back door we might eventually get more access. But really, it's crazy for our government not to be funding more research into things to do with our ocean. This is the life support system of the planet that we need to sustain us well into the future. To just ignore it is utter foolishness.

Over the past ten years we've seen a dramatic **loss** of funding for marine science. It's very expensive research: you need ships and submersibles and crews and fuel. There have been many decisions made that a lot of us are concerned about. There was a tremendous amount of money put into a high-end program called the *Integrated Ocean Observing System* that sounds good in terms of wiring the ocean, but in fact it's not very innovative and it's taking a huge amount of money to maintain the infrastructure.

This means that they've cut back on the mid-range funding to cover that higher-end cost. So now the grants that used to be available for $200,000 are going to be cut back to $30,000 or something like that. Well, you can't do field research with those small amounts of money.

Meanwhile, there have been cutbacks to NOAA for some of their undersea research. One of their most successful programs was the *Ocean Exploration Program*, which has been cut back to almost nothing.

HB: So you need to partner with James Cameron or someone like that....

EW: Well, I do think that in some ways we're heading back towards the world of the Medici where we're going to be dependent on these wealthy people who have big yachts and interest in the ocean. That's the only way we're going to be able to do this type of research—for the time being, anyway.

Questions for Discussion:

1. *To what extent do you think it is widely recognized how many forms of marine life are presently unknown to us?*

2. *In what ways is it justified to be sceptical in awarding funding for projects that are naturally unable to demonstrate what they will discover?*

3. *Does the state of being dependent on wealthy philanthropists for a public good represent a failure of politics?*

4. *Is Edie being too America-centric in her view of how the public sector can help safeguard the oceans? Are there international organizations and collaborations that offer hope for the future?*

IV. Grappling with a Coastline Crisis

Measuring toxicity and creating pollution maps

HB: Let me return to the coastal environment which you had mentioned earlier. You were talking about how trying to develop a way to sufficiently monitor the local environment here in the Indian Lagoon was itself an enormous challenge. Let's get back to that and tell me what sort of a role bioluminescence could play in all of this.

EW: When these reports came out—the Pew Oceans Commission, the US Commission on Ocean Policy report—they were the triggers that led me to start ORCA—Ocean Research and Conversation Association—a not-for-profit organization. Our focus is on developing technological solutions to ocean conservation challenges. One of the first projects we started up was a water-quality monitoring device called Kilroy that was developed by Eric Thosteson, a very talented engineer.

He designed Kilroy the way you'd design a cellphone: multiple sensors, one power supply, one communication system. He found a way to package it so that it was as inexpensive and user-friendly as possible. So we created these systems that we could put out in the water and attached them to existing dock pilings or channel markers out here.

We also developed a sensor package that we can put on it to measure bioluminescence. In other words, we can use that to look at the real-time biology of what's going on out there.

One of the cool things about all this is that the Kilroy can actually identify the luminescent animal to some extent by the type of flash it produces.

There is one particular type of dinoflagellate that produces something called paralytic shellfish poisoning, and they all produce the same type of flash. So we can recognize it by the type of flash that it produces.

HB: Paralytic shellfish poisoning? That sounds dangerous.

EW: It's very nasty. That's when you have a bloom of these red-tide organisms and you're warned not to eat oysters or clams. Obviously having a proper monitoring capability to be able to see that coming has a huge economic impact on the shellfish industry and could be very valuable.

So anyway, we developed these Kilroys. When I started ORCA we had kind of a three-legged stool of funding: federal, state and private. We ended up losing the state funding because the State of Florida went bankrupt; the federal funding has been gradually drying up in general for anything to do with conservation, and so we were dependent on private funds.

We were supported by a wonderful family foundation: the Claneil Foundation. When I made a presentation to their Board about the Kilroys and the progress we'd made, one member of the Board turned to me and asked how we knew exactly where to put the Kilroys. I really didn't have a good answer for him. I said something about looking at water-flow patterns, but the more I thought about it the more I realized that this was a really good question because you want to be smart about this.

What you'd really like to know first is where the pollution has been accumulating in the sediments as an indication of how to best map your flow patterns. Sediment sampling is easy enough, but the problem is that it's very expensive if you don't know what the toxicant is that you're looking for. You test for many different things, it takes a really long time, and it's very expensive.

So I thought to myself that what we needed was the equivalent of a canary in the coal mine. Coal miners used to take canaries down with them into the mines before they had detectors for the poisonous gases they had to worry about, because they knew that if they saw the canary suddenly keel over or stop singing they needed to get out of there quickly.

For us that boils down to what's called a "broad-spectrum bioassay": a way of using living organisms to test for a whole range of things they would be sensitive to.

I did a survey of different types of assays, and there was one using bioluminescent bacteria that obviously attracted my attention, because of my background. I didn't invent it: it's called Microtox and it's been around for years for testing food safety issues.

It's been used a little bit in Europe for sediment testing, but it's never really been standardized. We have a scientist here, Beth Falls, who standardized that assay and made it very, very dependable for us to be able to broadly measure the toxicity of sediments.

Now we use this as a trick to be able to make pollution visible. We go out, take our sediment samples, code them, and we end up producing something that looks just like a weather map, but instead of red being hot and blue being cold, we create pollution maps where red is toxic and blue is non-toxic.

HB: So let me and see if I've got this straight. The assay consists of collecting a bunch of soil and combining it with this bioluminescent bacteria. And if everything is fine, the bioluminescent bacteria just keeps doing its thing and emitting light as usual. But if it's in trouble because of pollution—if it's this canary which is about to keel over—it stops emitting light in the same way.

EW: Yes. In fact, what you're really measuring is how much of the sediment—the stuff off the bottom of the ocean—is necessary to make the light from the bacteria diminish to a certain amount.

HB: Which gives you a sense of the intensity or concentration of the toxicity in the sediment.

EW: Yes. It's called the EC50—the effective concentration of sediment to get 50% of the light. So if it only takes a tiny amount of sediment to do that, then that sediment is really, really toxic. And that, in turn, gives you your pollution map.

Once you have your map, it may be very obvious where the pollution is coming from—because, say, there's a bright red spot right in front of a boatyard or something that's been dumping heavy metals—but sometimes it's not so clear and then we put the Kilroys out to track the flow patterns so that we can try to understand where, precisely, the pollution is coming from.

We've made good progress with this, but it's been extremely difficult to get this funded as we go forward because of the lack of state and federal support for water-quality monitoring. As I told you a few moments ago, we were going to use state and federally-mandated water-quality monitoring programs as our funding stream. But they're gone now, and it's kind of amazing because most of the public is completely unaware that nobody is minding the store.

HB: Nobody?

EW: Nobody is monitoring this. We've got a situation right out here in this lagoon where we've had a catastrophic loss of seagrass beds within the last year and a half. Seagrass beds are like rain forests, and they're something like the third most biologically diverse ecosystem on the planet behind coral reefs and rain forests.

These seagrass beds are absolutely essential for life in these estuaries. They support these hiding places for the baby fish—again, it's all about hide and seek—even offshore fish come in to breed in these estuaries because of these hiding places in the mangroves and the seagrass. We've lost 36,000 acres of seagrass over the past year and a half, or about half, of the 156 mile-long Indian River Lagoon.

HB: So what's going on?

EW: Well, that's an interesting question, because it's been controversial. There was a microalgae bloom that was initially thought responsible for it because the microalgae block out the sunlight. But whether it lasted long enough to really kill the seagrass as thoroughly as it has—because the roots of the seagrass are gone—is far from clear. And in any event, that bloom did not extend down as far as the die-off did, so the two don't seem to coincide.

Questions for Discussion:

1. How can we be certain that any pollution-related assay is evaluating all relevant toxins in the water? Might there be some bioluminescent bacteria that are resistant to substances that are dangerous to humans? How might that be taken into account?

2. Are you shocked at the idea that state and federally-mandated water-quality monitoring programs no longer have a funding stream to support the kinds of activities that Edie describes in this chapter?

3. Why do you think it's important to independently evaluate both the sediment and the flow patterns to best assess and track pollution? Under what circumstances might the two not clearly overlap?

V. Kids to the Rescue

A reason for optimism

EW: Because of all of this, we've started a program working with school kids. As I become more and more of a conservationist and begin to take a 30,000-foot view of things, I realize that the only thing that is more unconscionable than destroying these ecosystems as we have is handing ecosystems that are spiralling out of control to the next generation without giving them the tools to deal with the situation.

When I was born there were 2.5 billion people on the planet. There are 7 billion now. That's just mind-boggling. All these people are using resources and creating waste and burning fossil fuels, and we are creating a chemical Armageddon as a consequence.

The fact that we could actually be *acidifying* our oceans is mind-boggling. And the toxins that are pouring into our waterways and poisoning ourselves as well as all of these ecosystems—how are these kids supposed to deal with this?

But this is actually a problem on many, many different levels. All conservationists ever seem to do is just preach doom and gloom, and it's been said many times that Martin Luther King Jr. did not mobilize the civil rights movement by preaching, *I have a nightmare.* But that's precisely what the conservation movement is trying to do—

HB: Well, you were pretty good at it just now, by the way. You did a great job of scaring me.

EW: I know, I know: it's a horrific situation. But I think the first thing we have to do with these kids is teach them optimism. And the way you do that is to do it *realistically*, because blind optimism is not the solution.

They have to know that they are problem solvers. I think this all harkens back to the way we teach science in our schools—which is very, very badly.

Some years ago, UC Berkeley child psychologist Alison Gopnik wrote a famous essay in *The New York Times* where she asked, "*What if we taught baseball the way we teach science? If we did that, then kids wouldn't actually get to **play** baseball. They'd learn the rules of baseball and when they got to college they might get to reenact certain famous baseball games, but it wouldn't be until they got to graduate school that they'd actually get to play a game. What kind of baseball players would they be then?*"

Well, that's exactly how we *do* teach science. I mean, we actually teach science by having kids watch a candle burn and write down their observations. Now, I know that a really gifted teacher might be able to make that work, but honestly that's about as exciting as watching paint dry.

We've got kids at an age when they want to make a difference in the world. They're passionate. Let's harness that passion for real change in the world. Let's get them involved in doing these measurements that nobody else is doing, solving problems that nobody else is solving—learning *how* to solve problems.

So we started a program that we call *Save the Water Babies*. The name comes back to the number of times that I have had some fresh young face sitting across my desk telling me that they want to grow up to be a marine biologist, and when I ask why, they tell me that it's because they love dolphins and they want to swim with the dolphins.

So I tell them: "*Look, if you really love dolphins the most important thing you can do for a dolphin right now is clean up its water.*"

I have pictures of dolphins in the southern part of the Indian River Lagoon here that are covered with this flesh eating fungal infection that is just horrific. If you know anything about dolphins, you know that they have extremely sensitive skin and they must be in agony. So if you want to make a difference, let's talk about the dolphins and the number of baby dolphins that are dying due to the toxins in the water because the mother passes on her toxin load to her first-born through her milk and blood. That's why we call it *Save the Water Babies*—we can use lots of pictures of cute baby dolphins. And kids get it.

We then got this wonderful grant from a very interesting community organization called *Impact 100*. The idea here is that one hundred women in a community get together and each of them gives $1000 to create a $100,000 grant that can be given to any one organization to create a true impact.

We're using this grant to work with Indian River Charter High School students to have them create a pollution map. They've gone out and created a pollution map in the northern part of Vero Beach where we've lost most of this seagrass.

We haven't reported this yet because we've just started to get the data back, but the toxicity in the sediment is the highest we've ever seen anywhere, and it's not what I was expecting because that's not an industrial area. I don't even know at this point *why* it's so toxic, but these kids are incredibly motivated now.

They're connecting the dots. They understand. For the first semester, they're collecting the data and creating the pollution map. Then for the second semester, based on what we discover, they're going to create messaging for the community and try to develop public service announcements and whatever other ways they can think of that can make a difference.

Many kids applied for this class—we couldn't take them all—and we tried to get a balance. We wanted kids that were good at art as well as science because we wanted to create as diverse a group as we could to make as powerful a team as we could, in order to address these problems and show these kids that they can make a difference.

HB: Just to highlight this notion that Alison Gopnik was writing about, you're getting high school students to engage in *real* science—not just hear about or talk about something, not just do it as an experiment because they're told to do it, with everyone knowing ahead of time what *should* be found—they're really engaged in the process of doing scientific research. Which is impressive enough. But they're doing something even *more*: they're engaged in helping to provide solutions to an issue that's of direct concern to *all* of us.

But as I'm listening to you, I'm thinking, *Can we somehow scale this? How can we make this bigger than the Indian River Lagoon?*

That's obviously of enormous import to people who live here, but this is something that we should be able to do on a regional level, a national level, an international level.

There are high school students and teachers all over the world who I'm sure would be very interested in doing something similar. Is there a way we can scale your particular program? Are there other programs that we can somehow manage to put within this rubric? Because it seems like such an important idea.

EW: I think it can be made scalable. That's been a focus of mine recently: how can we make it scalable? I'm not sure that we can do it with the Microtox assay I was just talking about because that requires some fairly expensive equipment and probably more expertise than your average high school group is going to be able to bring to bear, so we're trying to develop some simpler lower-cost assays. My ultimate goal is to actually make these available on the web as open-source material and then have other people feed into it. There are many citizen-science groups where kids take on civic responsibility issues. There are other groups that could come to this from all different angles.

If we put it out there so that these kids can do this themselves, then absolutely I think it could be scalable. The trick right now is to find a bioassay that is simple enough and dependable enough and robust enough that this is going to work. We have some research we have to do before we're going to be quite there, but we've been making some real progress on it.

HB: In terms of future support, it seems that one of your obvious advantages is the appeal that this program has across such a broad spectrum: the educational community, the philanthropic community, the environmental protection community and so forth. Do you have a sense that there is any momentum building in terms of funding and general awareness at a national level?

EW: It's been a very hard slog but I do feel like we're finally getting some momentum. Our efforts are now starting to get recognized—these pollution maps really make people sit up and take notice.

HB: Well, they see something concrete that is detailed and scientific.

EW: Yes. I'm actually more optimistic than I've been for quite some time about this because I've been struggling for a really, really long time to try to find a foothold to get these concepts across.

But if we can harness the energy of the youth to solve real-world problems, that may be the tipping point for a whole lot of different things.

Questions for Discussion:

1. Do you think that the environmental movement has spent too much time scaring people about the severity of the issues and not enough focusing on possible solutions?

2. In what ways would science education be improved if practicing scientists had more of a role to play in the classroom? How could their participation be more broadly integrated in a scalable way?

3. What are the main factors stopping high school children from more direct, laboratory-based activities? Those with a particular interest in this issue are referred to Chapter 10 of the Ideas Roadshow conversation **Indiana Steinhardt and the Quest for Quasicrystals** *with Princeton University physicist Paul Steinhardt.*

4. Have you been involved in any citizen-science efforts in the past?

VI. Existential Challenges

Global concerns and worthy precedents

HB: Moving from the very difficult to the perhaps even more difficult, we've talked about the educational and ecological issues and the spectacularly poor situation locally that you unfortunately seem to find yourselves in, but presumably what's happening in the Indian River Lagoon is not as unique as we might think. Looking at matters globally, you mentioned that we only know 5% of what's in the world's oceans; and you ask pointedly: *How on Earth can we best protect something if we don't even know what's in it that needs protecting?*

EW: Yes. I should have pointed out that the year the US Commission Ocean Policy report came out was the same year that we discovered that squid that nobody had ever seen before. And this just brought home the message that we're destroying the ocean before we even know what's in it.

That's what's directed me towards this *Save the Water Babies* program. Because we don't have a populace that has a high enough understanding of the basics of ecology and ecological systems and how they support our existence on the planet. And I think the complexity of those things has deterred students from learning about them early on, and certainly later on as adults.

So to get them actually out there in their local ecosystems making a difference for their local communities is a great way to end up with a much more educated populace that can better understand what, after all, ought to be a simple concept: that if you have a marine reserve and a no-fishing zone that covers 10% of your coast line, you're actually going to have 50% more fish that are possible to catch in the longer term. But short-term greed overtakes long-term gain again and again and again.

We have to involve people who understand those basic principles and can fight for them, people who have been out there and have seen for themselves what a difference small changes can actually make. Because our individual choices—what we eat, what we put on our lawns, what kind of cars we drive—all of these things are impacting our life support systems.

HB: Do you see reasons for optimism? Are attitudes changing? You travel around and do a lot of talks and have many discussions with people of all different orientations and persuasions. Have you noticed a growing change in perspective?

EW: I think there's a lot of reason for optimism. For one thing, we have more knowledge at our fingertips than ever before in human history.

The other thing is that one of the real points of greatness of this country in particular is that we're still a young country and we think like a young country, which means we can actually turn on a dime.

We've done it again and again. Maybe the best example of that is the campaign to stop smoking. I know that for some people it seems like it went on for a long time, but eventually truth won out, and as far as the rest of the world is concerned, the US stopped on a dime—smoking is just not publicly acceptable here anymore, and that happened in an incredibly short period of time. For something that was such a cultural centerpiece of every place that you went, to now have it be so vilified because of health was remarkable. People get it.

And it was largely the youth that turned that around when the tobacco companies were fighting some of the public service announcements. The ones they went after the strongest were the ones that were being produced by young people because they were too effective. They were fine with the ones that the Cancer Institute was producing, but the ones that the young people were putting out for other young people were too persuasive. So I think the young people can make a huge difference, and I think they are our real reason for optimism.

HB: Let me conclude by asking you one more question. I know that you're a scientist and an activist and not a political person. But suppose for a minute that you had just been elected President of the United States. What would you do differently? What would you do to turn all of this around?

EW: It would all be about education. I would just focus so hard on education and the difference that it makes: that's the economic engine of this country.

Education about the environment would hopefully lead to more funding for conservation efforts. All of those things would just fall in line if we could just put our focus on what has been the strength of the country for so long and unfortunately no longer is one of our core strengths: innovative education.

The most Nobel Prizes in the world have been awarded to American scientists, which is a reflection of our basic innovative thinking. That's still there. Let's not lose it. Let's tap into it and put our resources into as much innovation as we possibly can to work together to solve all these problems.

HB: Well Edie, listening to you passionately and eloquently describe your views has certainly been a very educational experience for me. Thank you very much for allowing me into ORCA, your home by the sea, and the very best of luck with all of your very worthy and important initiatives.

EW: Thank you.

Questions for Discussion:

1. *Should intergovernmental groups such as the United Nations play a greater role in regulating ocean policy? What are the advantages and dangers of such an approach?*

2. *Does this conversation make you more optimistic, or more pessimistic, about our ability to deal with the despoilment of our oceans and coastlines?*

3. *Does the media do a good enough job in involving people like Edie to spread their message? How might that be improved upon?*

4. *Do you find Edie's example of how public attitudes towards smoking rapidly shifted appropriate? How might we be able to harness lessons learned from that experience? For a related perspective on this issue, see sleep scientist Matthew Walker's similar invocation of changing societal attitudes to smoking in Chapter 10 of the Ideas Roadshow conversation **Sleep Insights**.*

5. *What would you do differently to address the issues discussed here if you found yourself in an influential position in government or a well-funded and potentially impactful organization?*

Continuing the Conversation

Those who enjoyed this conversation are strongly recommended to read Edie's captivating memoir, *Below the Edge of Darkness*.

Coral Reefs

Science and Survival

A conversation with Charles Sheppard

Introduction

All Too Relevant

A standard problem encountered by those who opt for a career in the mathematical sciences is the often significant disconnect between their work and everyday life. There is, undeniably, a gratifying level of eye-raising respect that you get when you tell people at a party that you spend your days grappling with scenarios of what the universe is doing in the first few microseconds after the Big Bang, but after the initial impression fades there is a certain lingering discomfort in being forced to admit that the doctors and engineers and entrepreneurs around you are engaged with the hustle and bustle of "the real world" in a way that you can only barely imagine.

Of course, such independence from the "crooked timber of humanity" has its definite advantages, and virtually everyone who has felt the pull of the "hard sciences" recognizes that its very remove from "the human element" is, in fact, a large part of the attraction. But still and all, there are times when every physicist or mathematician, usually after weeks of fruitless calculations and a drink or two, looks herself in the mirror and asks, *Is this actually relevant to anyone? What difference does any of this really make, after all?*

For environmental scientists, meanwhile, the problems typically lie in precisely the other direction. The issues at stake are overwhelmingly relevant to everyone—indeed at times existentially so—but finding a genuine solution isn't even half the battle. Sometimes it is much, much, less than that.

Charles Sheppard is a specialist on coral reefs, which are not only one of the most spectacularly diverse ecosystems on the planet, they are also the primary food and survival sources for millions of people around the globe. And they are dying. Fast.

> *"If you think of all the reefs of the world, in rough proportions: about a quarter are dead; about a quarter are very badly damaged and at the present rate, are going to be dead in not too much time; a quarter are damaged a bit, but they would recover if they're allowed to, and only a quarter are left alive.*

> *"Now, we turn to the dead and damaged three-quarters, and ask, **What caused that?** We can look and see what impacts they've had. In many parts of the world, the impacts are very, very severe."*

You might be tempted to conclude—as I initially did—that identifying the root causes is a complex problem, or that any putative solution would be prohibitively expensive or impractical. But you'd be dead wrong.

> *"Essentially, we know what to do to maximize the production—the surplus protein, if you like—which you can take from a reef. We know that. We know how not to kill a coral reef. We know that. It's a matter of convincing authorities, what they call the decision-makers, 'important stakeholders'—to use that horrible phrase—to do something about it and to do what you say.*

> *"Time and time again, I said, 'You haven't got this source of protein for these villages, that's why the villages have been evacuated, because you've killed them by X, Y, or Z.'*

> *"And the minister of the government might say something like, 'Oh, no one told me that was important.' Well, we have till the cows come home, but we can't get through enough. There's this gap between informing the decision-maker, who might or might not be ruthless himself; he might be quite benign and have the best interest of his people at heart. But he hasn't had the information that people like myself have been shouting about for a long, long time."*

Why? Well, Charles believes that many factors are responsible for our current state, from a lazy and often unresponsive media to short-sighted politicians to predatory multinationals who are indifferent to the long-term effects of any of their policies on a population that they will soon move away from.

As if that isn't bad enough, often those who should be aligned with progressive environmental policies can be just as obstreperous and counterproductive. The creation of the Chagos Marine Protected Area in the Indian Ocean, for example, is viewed by many as one of the few bright spots in the battle to preserve coral reefs. Established in 2010 within the British Indian Ocean Territory, it is one of the largest marine protected areas in the world. But there is also a joint UK/US military base in the region, which for some people necessarily damns the entire venture.

> "The opposition to the Chagos Marine Protected Area comes from a number of sources. Very often, some of the most vitriolic that I have faced—because I was one of the people working in the background to help create it—has come from people who just hate the idea of the American military, or the Western military. So, in their view anything done there is going to be wrong to start with. Some people have told me that outright: 'Well, nothing you do is going to be right. And I'm associated with 'a capitalist, imperialist machine'.'"

What's a dedicated, pragmatic, environmental scientist to do? Well, Charles believes that there are some genuine, positive steps that can be envisioned despite the slings and arrows on all sides.

> "My main goal would be to see a marine spatial planning system, which includes areas that have already been sacrificed—completely covered with fish farms, say—through to multi-use and right up to "no take" protected areas: the whole suite, rather like Western countries do with agriculture now. You have your national parks; you have an area which is a farm, with almost no diversity to speak of other than corn. And you'll have some areas where you can't do anything. I think that's the type of solution we've got to translate from land to sea."

But even this, Charles realizes all too well, is far easier said than done in a world where all too often nobody has any clearly recognized jurisdiction over the ocean and most commercial activities are permitted unless it can be unequivocally demonstrated that they will cause harm—an almost impossible standard.

> *"I've been in many situations when I've heard a company say, 'Well, there's no proof it'll do harm,' and I replied, 'Well, up the coast there, we know that a similar thing has done harm.' And they'll just say, 'No, that's not proof it's going to happen here.'*

> *"That's completely fatuous. We **know** what will happen if they do that. The burden of proof must be turned around, saying that the mechanisms they're going to use to do their development won't cause harm.*

> *"Now, I should say that this development in question might bring benefit to huge numbers of people. I'm not anti-development—people often make that easy mistake. But there are better and worse ways to do things, and the better way might sometimes be more expensive than what the company had in mind at first."*

Despite the best efforts of his many determined opponents to portray him as such, Charles Sheppard is neither an anti-development militant nor an integral cog in "the capitalist, imperialist machine". He is, instead, a dedicated, hardworking scientist determined to try to find a way through political shortsightedness, corporate greed and societal indifference to use his hard-earned expertise to make the planet a better place. But it certainly hasn't been easy.

> *"People going into the line of work I've been in—whether it's land or sea—need to know the science, but they also need to know quite a lot, as much as they can, about how systems work in the countries they're going to. They've also got to be robust in themselves, because the vested interests can be fairly powerful and very strong, and you get a lot of ad hominem attacks. A lot of people I know have been put off completely by doing that. But hey, you keep on."*

You think cosmology is hard? Try doing something "relevant".

The Conversation

I. Watery Beginnings

The power of scuba diving

HB: I imagine that when you began your research career, the lay of the land in terms of environmental science was very different than it is today.

CS: Yes, that's right. When I first began my career in marine science and environmental science—it's not only reefs, it's impacts on affected areas in all the world oceans—that was in the early '70s, and there were no desperately bad bits, and kind of good bits. Well, there were some desperately bad bits in some of the megalopolises in the tropics, but they were quite small. There were lots of very healthy ecosystem around; and those that are the best bits now, because they're protected, were no different to those that existed everywhere else.

So, what has happened is that the bad bits have gone downhill, the gap between what's left and the bad bits is increasing significantly, the amount of good bits are shrinking quite rapidly and quite alarmingly, and only about a quarter are in good condition still.

The prognosis, with the momentum we're talking about, is not very good. Another issue is acidification. That's got a 30-year lag for the CO_2 in the atmosphere to reach equilibrium with the CO_2 that is dissolved in the oceans to cause the acidity. So, even if the CO_2 didn't increase anymore now, the acidity of the oceans is going to get worse for the next 25-30 years anyway. Whatever we do, we've got to bring it right down.

So, there are these, what we call, "legacies", which have a built-in time spread.

HB: And that must be difficult for the public to understand, because I think people are used to cause and effect happening on much shorter time scales.

CS: It is. Most of the marine impacts around the world, around our oceans, are short term. If you build a lead acid factory in an estuary, you won't have a very long lag at all before the estuary is wiped out, say. I've seen this in some parts, and the people have to evacuate. If you fill in a nursery, a fish breeding nursery ground, because you want a landfill to build on, you get severe impacts: there are fishing villages in Southeast Asia—I've seen them—which are abandoned because the fish aren't there. That's a fairly short time scale, it still takes people by surprise.

With global warming, a warming pulse has an effect, which is slightly longer, but things like acidification have a very long lag time, which are quite out of our normal expectation and understanding of what a cause and effect is. And that is one of the problems; but even without acidification, the downhill slide of a lot of ecosystems is happening at a terribly fast rate. And as resources get shorter, people use more and more desperate means to try and feed themselves. In Indonesia, for example, dynamite fisherman on reefs can earn two or three times as much as a university professor there.

Now, whatever your views of professors, this just shows how quickly habitats can be destroyed. You don't get the fish back for generations, and the local people are left to pick up the pieces as it were; except in cases like that, there are no pieces to pick up—there's certainly no fish to pick up. And that happens again and again. In West Africa, I've recently heard about a case of someone fishing using DDT. That's a pesticide that we've banned in a lot of countries in the West, but there are bags of it in West Africa; he'd pick a standard handful and throw it in the water, and any fish that died and floated to the top, he could feed his family on.

And when he was told, "*You know, that's very bad for you,*" he just replied, "*Well, I know that. I know, in the end, it's probably going to kill me; but if I don't eat now, my family's going to die this week.*"

So, I think one of the problems we have is what I call "a comprehension gap". Many people in the developed world just can't conceive of that kind of hardship, and that's the first hurdle to overcome. I remember a very senior person in government once saying to me, as I was talking about the interconnectedness of the UK, and places in the tropics and what the problems were, "*Well, that's a long way southeast of Dover.*"

And you can be aghast at that, but you can also understand that it's the people who are not southeast of Dover that reelect him. And I said, this was a senior person, so if the comprehension is lacking there...

We need to try to get the message across that it's increasingly an interconnected world. And if other countries are to collapse, it *does* have a consequence on us. We can see that in the politics of the news any day this week.

HB: When you began your research, were you always thinking about the impacts, man-made or otherwise, on these marine ecosystems? Or were you just transfixed by the ecosystems themselves? Did you just think, *This is fascinating stuff, I'd really like to study it, find out how it works.* Or was it more a question of the impacts on those ecosystems?

CS: My PhD was about impacts around the UK: we were measuring ecosystem changes as you went into pollution gradients, into polluted areas. Then I got the opportunity to work on reefs, and I had a postdoc in Australia, where it was pure ecology. It was at the Australian Institute of Marine Science and was the best sort of opportunity you could have. They essentially said, "*We're going to pay you for two years, you can do what you like.*"

They expected some results, of course, but I've been trying to get that sort of arrangement ever since. So, that was pure ecology, but after that I began looking at reefs, focusing on the impacts and monitoring—not only reefs, but mostly reefs—in the Mediterranean and other areas.

HB: And had you been scuba diving or snorkelling as a small child, exposed to the sea more than most?

CS: Very much so. I was born and brought up in Singapore, which was more marine orientated in many ways. I was offered a PhD after graduating in medical research   I was doing a medical physiology and pharmacology degree—but I began scuba diving as a hobby and began to wonder, *What's all this around me? This is really cool stuff.*

And I was very, very fortunate. After getting my undergraduate degree, I turned down that offer and instead took up an offer for a doctorate at Durham with someone called Professor David Bellamy, who offered me many wonderful opportunities.

HB: And this was because of your scuba diving experience? This was the primary motivation to change directions?

CS: Yes. I began to go scuba diving as a hobby around Britain and was completely captivated by the whole experience and by what I saw.

HB: Where does one go scuba diving around Britain?

CS: Well, the best parts of Britain to go scuba diving, I think, are the West coast because the water's clearer. Some parts of the East coast are good as well, but it can sometimes be very muddy and not so interesting from the diving point of view. But on the West coast, the water can have the clarity of the tropics.

There are also very interesting types of seaweed. We have a kelp plant here, which is about the height of a diver. It's just a magical place. In the summer, with clear water, and the sun shining, it's really good. It's cold. But it can be glorious.

HB: So, your scuba diving experience motivated you to move into a different research direction. In terms of your motivation towards environmental issues, did this become awakened in you as you progressed further in your research? Or was this also something that you had from an early age?

CS: It was part of me from an early age, because I was born and brought up in the far East. I could see the differences between myself and the European kind of lifestyle, as an ex-patriot—even as a child, you can see it quite clearly—and the local people. The differences were stark and that struck me then, and it was glaringly obvious.

I've long been fascinated by the idea of research. I really enjoy that, I always have. I like writing as well, and they work together very well; and I did have the opportunities. After getting a postdoc, I went to work in the Middle East and that was quite eye-opening, the difference in attitudes towards nature.

HB: Where were you exactly?

CS: Well, my first position was in Saudi Arabia. That was on the Red Sea, the development of a big oil terminal there, which was going to be built on reefs and things like that. And I was struck by the lack of awareness of what the damage could be, much of it was avoidable.

OK, you're going to have an oil terminal, that was never up for debate. But there are ways to do it where you can have your terminal and not destroy the reef system there, or you can have it and maintain the reef system there, and it needn't cost you a lot more money, either. You can have your cake and eat it too. That was what my first job was, and that became interesting. It's naturally wrapped up in economics, because people will say, "*Oh, it will cost too much*." Well actually, it needn't.

And when you do the proper costings, the all-in kind of costings, you realize that when you damage things and have to repair it, or you suffer because you haven't repaired it, it's much more expensive in the long run. The economics of ecological damage are something, which is very much undeveloped, I think.

HB: As you were talking, I thought of a medical analogy, or at least an argument, which is often put forward in terms of prophylactic medical care: that if you spend the money to ensure that people will see a doctor regularly—which certainly costs quite a fair amount—at the end of the day it costs a lot less than if you don't do that and people wind up needing all sorts of advanced medical care.

CS: Yes. I use medical analogies quite often when it comes to impacts on marine ecosystems: if you had the flu, malaria, and whooping cough all at the same time, and three others as well, you probably aren't going to survive it. But most individuals can quite easily survive all of them one at a time.

So, if you have impacts on a reef system, say, which include over-fishing, sewage, industrial pollution, landfill, and things like that, all at once—which happens in a lot of the world—then the reef just can't recover. But, if you have any one, or another, they might survive that, at least for a limited length of time.

The other analogy with medicine is that prevention is so much cheaper than cure. When it comes to reefs, we don't know how to cure them, really. If a reef is dead, as about a quarter of the world's reefs are, we don't know how, really, to help them come back to life again, other than simply leaving them alone, and that might not be an option.

Ecology and economics have the same root, *eco*. And they should be two halves of the same coin, but they're not: they're just so far apart now. We can't do ecology because of economics far too often, yet we can't have the ecology, which we depend on, without the economics. So, it's just a pity that they have diverged from each other so profoundly, and the one is tied up with the other.

Questions for Discussion:

1. Should activities like snorkelling and scuba diving be explicitly encouraged in schools?

2. What is the precise meaning of the Greek root "eco" that Charles refers to here?

II. Building A Reef

Coral, algae and time

HB: I'd naturally like to return to that question later, asking very concretely how we might be able to move forwards towards achieving a greater level of coherence between ecology and economics. But first, I'd like to talk in more detail about the reefs themselves, and I'd like to start at a very basic level, which is what are these things anyway? What is a coral reef, actually?

CS: Well, a coral reef is built by corals, stony algae, and some other things as well. Corals secrete limestone as a skeleton—

HB: OK, but allow me to interject for just one moment, as I'm determined to start at the very basics. What is a coral, exactly? Is it a plant? An animal?

CS: The coral is an animal, which is like an anemone or a jellyfish, but upside down. It secretes a limestone cup around itself, which is its home, if you like. And as they build their limestone, they grow upwards slowly. The reef they build as a result of that is thousands of years worth of accumulated growth. Not all the coral that grows adds onto the reef, a lot becomes washed away.

HB: What percentage would that be, roughly?

CS: Coral can grow between a centimetre and 10 centimetres a year, depending on the species. A reef, itself, will grow probably 5-8 millimetres a year, if it's in healthy condition. That's because a lot of things erode it away at the same time that it's growing: wave action, parrot fish, sea urchins, which scrape it, and a lot of boring organisms, which make holes in it and it becomes like Swiss cheese. So, it disintegrates and becomes sand, which ends up on a beach, which builds a coral island. So, reefs can be a mile or two thick. The deepest parts were deposited millions of years ago. So, over that time they've grown up.

Now, they're marine animals, so they cannot grow above the low water mark. They will stop at the low water mark. So, reefs all around the world reach the low water mark at the highest. And there's a constant battle between erosion and growth: wave action, and the animals and plants I just mentioned, erode it away. And you have the growth of the corals, which are secreting their skeletons.

Now, I said a coral is an animal. But the reef-building species contain algae, single-celled algae in a symbiosis. And like any algae, just like any plant, they photosynthesize. And anything up to 80% of the energy from a coral comes from photosynthesis.

HB: 80%?

CS: It can be that much—for some. Not all of them have the algae, they are entirely carnivores and they will eat the zooplankton from the water that they capture using the many stinging cells they have on their tentacles. Any coral will do that and get energy from that process, but a very high proportion of their energy comes from the algae, which are captive in their cells. As I said, it's a symbiosis. Those algae also help them to grow faster too. And it's only those that have that symbiosis that really contribute to the growth of the coral reef.

HB: I see. And so what percentage of those that have that algae would there be, compared to the other non-algae containing corals?

CS: As far as we know now, it's about even: about half the corals have the algae symbiosis; they live in the tropics and are limited to warm and clear water. And the other half—I bet there are more than half, actually, but we haven't found them yet—can go right down into deep, dark, cold waters and right up to the poles as well; they're not dependent on light, and they're entirely plankton feeders. So at least half the species are that. They're small, usually, slow growing, but they can form big reefs down in the Abyssal Depths, as well.

HB: Oh. They can?

CS: Some can, some can. Most of them are small, sort of cups, which you see all over the place on the rock, not conspicuous so much. They don't form reefs like the tropical reefs that we're talking about now.

HB: As you're talking I'm wondering, I have these two different species, one can form a reef, one can't, I'm wondering about what the evolutionary advantage might be of one with respect to the other. Presumably having a reef around is a boon for your entire ecosystem: these things are some of the richest areas of biodiversity on the planet, so I imagine that there is a clear form of "community good" in operation going on there that helps a whole lot of species. But if I just look at the situation in terms of one type of coral versus another, how does reef building actually give an evolutionary advantage?

CS: It must do. It's not only corals that have a symbiosis. Going way back in geological time, reefs have been formed by quite a range of other organisms: from those that look like coral, but are not really related, to other organisms, which look like mollusks. Quite a number partake of this symbiosis, which confers on them an evolutionary advantage. It helps them to grow very much faster: they get this huge source of energy from the sunlight. They are limited, though: for some reason they can only live in warm waters, the tropics. Those that don't have the algae can live in deep dark areas, all the way up to the poles.

HB: So, they're more flexible in terms of location.

CS: Yes. They're very much more flexible in the range of places where they can live. They *can* live on a reef, but they're out-competed maybe. You see them on reefs. You see them in overhangs in caves, you see them. But they're never dominant.

HB: Interesting. I'd like to talk a little bit about the mechanics of reef formation. You mentioned in your *Very Short Introduction to Coral Reefs* that Charles Darwin wrote a seminal book on coral reefs.

CS: He did. I think that Darwin's ideas in his book on reefs, which he developed after his voyage around the world, were more original than his theory of evolution. The idea of evolution had been in the air a long time, they just hadn't thought how it could work. Now, I'm certainly not trying to diminish his achievements there, but given that it came from a brand new start, I think his book on reefs probably was the most original and innovative.

He thought—and it's been shown to be true—that the corals grow, as I said earlier, by depositing their limestone cups, skeletons, exoskeletons, growing upwards. And he formed a scheme where you could have an island with the corals growing around it, and the island would sink under its own weight. He had found, also in his trip around the world, that the Earth was much more "movable" than had been previously thought: he saw shells high up in the Andes and things like that.

And as the island sunk, the reef would keep growing up, because the corals were growing up towards the light—they can't grow higher than the low water mark, they would keep growing up to the surface of the water as the land sunk—and that would evolve into a fringing reef, then a barrier reef, and finally an atoll, when the land went completely.

And you can see around the tropics now, examples of every single stage of that—and a few more in some areas where you have a lot of tectonics, and the Earth jumps in steps.

He was essentially right, but there are additional nuances he didn't have: it was too early and he didn't see them. But in the next century, the idea of sea levels changing up and down as the ice ages come and go became certain. And the two together form our modern understanding of reef formation which is all related to present sea level. It's controlled by two things: the sea level per se—as in how much is or isn't in the ice caps—and movements of the land, because land in some parts of the world is being jolted upwards, while in others it's subsiding slowly, or perhaps being jolted downwards.

And the combination of all those factors makes a complex thing. But basically, the core of the whole story is that it's driven by the corals needing light; and they can grow upwards by depositing their own limestone. So, they form these reefs, which reach the low water mark.

HB: Does this mean that if I look at any reef or atoll, I should always find in the middle of it, some land mass, which has subsided, which is down below?

CS: In any atoll you would. Now, because of the time, and repeated changes of sea level, it isn't always so obvious to recognize—it's not so much like an atoll ring of corals indicates the original edge of the volcano—it's all got much more distorted. It can be a mile thick. The hard sea bed where the whole process began might be a mile deep. And over time, you've had all sorts of changes of sea level, and Earth shifting, so you won't necessarily find under one atoll, one volcano. You might. But, it's more of a messy story now, because of the length of time, the tens of millions of years, and the thickness that is now there.

In the 1950s they did deep drilling programs on atolls in the Pacific. They went down a long way—some were about a mile deep—before they struck the non-limestone rock that was the core of the whole thing. And that, essentially, proved Darwin right. I think people probably knew he was right by then anyway, but that was a very tangible sort of proof. Now, using seismic techniques, that's been done on a lot of islands.

HB: So, I get the basic picture that you start off with these animals that need to harness the sunlight so they need to be fairly close to the surface, and they start forming this fringing reef, doing their thing in terms of secreting this limestone exoskeleton. But I have a very basic question, which is, *How does the thing start?* Perhaps I'm missing something obvious, but I see how it can build one once it begins, but I have a harder time imagining how, exactly, it first starts.

CS: Well, there's a very good example of that in much of the Seychelles. A lot of the Seychelles are rocky, high islands; and you get the corals just growing on them, sticking on them.

The ocean is, or can be, or should be, a larval soup; and enough larvae are there to stick onto the rocks and they just start to form. I've seen some areas where, for other reasons, the corals are spaced out, and the rock between them isn't reef, it's not limestone.

Meanwhile, in other circumstances, they will grow and it will start to coalesce. I've seen that in Oman, as it happens, where the corals start to coalesce into a rudimentary reef; and then the process is off.

HB: How many types of corals are there?

CS: Worldwide, there's probably a thousand or so different species of corals.

HB: A thousand?

CS: That's not so much when you compare it with the numbers of fish, or numbers of insects, or something like that. But, there are about a thousand of reef-building species of corals. You'll get 800+ in the region that we call the Coral Triangle in Southeast Asia, and the numbers will reduce as you go away from that area.

That applies to the land, as well, to birds and insects on land. Southeast Asia is the real hot spot for diversity, tailing off east and west, and also north and south, from there. There are exceptions to that. In the Red Sea, there's another sort of hot spot with a lot of endemic species there, no one near as big, but by and large, you get a tailing off as you move away from the Coral Triangle.

And a lot of reef areas, in the Pacific for example, might only have 50-80 species. But their reefs are every bit as strong and large as the reefs which have been formed from hundreds of species. If you have a tailing off of diversity, the species that remain—that have got there by chance, maybe—just occupy a bigger proportion of the space.

Typically, in order for a coral reef to be healthy and still growing, it will have about 45–50% of coral cover—a reef will likely be struggling if it's much less than that—and in some areas that can occur from just a couple of species, actually. There are areas, because of environmental conditions and so on, where you have almost a monoculture of one sort of coral; and that will form the reef every bit as much as an area in Southeast Asia, or the Indian Ocean, where there are lots of species.

HB: And if you have a reef that's formed by only a small number of different species, what sort of effect, if any, does that have on the subsequent ecology and biodiversity of the entire reef system?

CS: One of the important things about a reef is that it forms a three-dimensional structure. So, if you have branching corals, for example— and about a third of them are—they form habitat for a lot of other creatures, as well. So, if you can form that structure, if the corals there are only a handful of species, but if they form a great three-dimensional matrix of habitats, then the ecology for other things might be just as good.

And that's the problem—well, one of the problems—that reefs face: a lot of environmental impacts can kill off the more delicate branching forms first and most easily. You'd need to think no further than dynamite fishing, for one thing. But also, warming tends to affect those more. And if you remove the three-dimensional structure of a reef—well, it's like chopping down the trees in a rainforest: then the birds and monkeys go too. If you chop down the marine rainforest, then the birds and monkeys of the sea go too, first. And by that I mean the fish, starfish, the shells—anything you like—all of them. So, maintaining the three-dimensional structure is what is important. And in some of the least diverse reefs of the world, the species are branching, so you've got that habitat still there.

HB: So, by "least diverse", you mean in terms of the spectrum of coral that are producing it?

CS: Yes, that's right. You can go to some areas of the tropics, the far eastern Pacific, or the Caribbean, actually—all of those areas are typically poor in diversity compared to the Indian and Pacific Oceans—but before they had damage as well, they would form very robust three-dimensional structures, which housed amazing ecosystems.

And I was fortunate enough to see some of those before the last three or four decades, when some of the branching ones have been very severely reduced.

HB: Obviously a principal theme for today's discussion is to look in detail at the environmental degradation, what is causing it and how we might be able to mitigate matters, but just to return to how reefs grow under normal circumstances, you said that we're generally talking a few millimetres per year, right? Is this what we're talking about?

CS: Well, the coral, which is a big dome, might expand its radius by up to a centimetre a year. A branching form can extend each of the fingers by, roughly, 10 centimetres a year.

But their limestone, which they've deposited, is more porous: it's got more holes, more space in it, you could actually dig your thumbnail into it for a lot of them. And they're more easily broken down and form a lot of the sand—which is an overlooked and equally important part of a reef ecosystem: it houses a lot of organisms in the sand, and all the coral islands of the world are totally dependent on that for their existence.

Waves will throw up the sand onto the shorelines, or make a pile of it, plants will settle on it, birds bring seeds onto it and so on. I don't know how many nation-states there are around the UN table, which are built wholly from coral sand and debris, and which are dependent wholly on the coral animal for their existence—but quite a lot. And a lot more states are partially dependent on them.

Some, like the Seychelles have atolls, which are dependent, and they have their high islands too—which will stay there even if all the corals go—at least the geology will stay there.

So, that's why, I think, some of the countries like the Maldives, Tuvalu, the leaders of those are really shouting quite loudly at the leaders of the western countries to get their act together when it comes to CO_2 and the consequent global warming, which is the immediate main global impact affecting reefs.

HB: As well as the impact of rising sea levels on the islands themselves.

CS: Yes, indeed. Some islands in the Pacific have become evacuated already. This is not a thing for the future, it's happening now.

Questions for Discussion:

1. What do you think the algae gain out of the symbiotic relationship they have with the reef-building coral?

2. To what extent were you aware of the importance of sand in the coral ecosystem before you read this chapter?

III. Gratuitously Unsustainable

The problem with humans

HB: I'd like to talk a little bit about the extent of biodiversity in a reef system. I've heard it said that a healthy reef system is one of the most biologically diverse environments in the world. Perhaps you could just flesh that out a bit more and describe the extent of its biodiversity.

CS: Well, the area of reefs around the world is well under 1% of the surface of the Earth, but they are the richest ecosystem in the world, I would maintain, in terms of diversity.

For a start, the sea is richer than the land at every level above the species and genus. There are many phyla of organisms, of which only a few live on land. If you take away the beetles, then they're the richest by any measure, really. As Darwin, I believe, said, "*God is inordinately fond of beetles*." If you take those away, the reefs have it.

They contain very high proportions of—well all of them—fish, of the big groups that you have in fresh water and the sea, reefs have just far, far more. They also, which is important in human terms, have some of the highest productivities of any ecosystem in the world. I understand that a Queensland sugar cane field, beats it, but that's only sustained by a huge application of fertilizer, and things like that.

HB: What do you mean by productivity here?

CS: The production of "top surplus" protein that people can eat.

You can measure production in terms of, let's say the amount of carbon fixed, which starts off by plants, of course. You have a secondary production on that, herbivores; and you have the carnivores eating the herbivores, and us eating the carnivores, say.

HB: So, biomass to some extent?

CS: Yes, the production of biomass. If you have a very high production of biomass, you can remove a lot. So, the standing crop might not be very high, and you see that on a coral reef.

If you were to cage a bit of reef and exclude all the herbivorous organisms—fish, sea urchins, things like that—in no time at all it would be clogged with seaweed, but it's just been cropped. So it's very fast rates of production with very fast rates of consumption. And if you do it right, which we rarely manage to do, you can cream off the top—eat it for food for people.

So, not only are they the most diverse system on the planet, they're one of the most, nearly the most, productive as well. And yet, they're so small in area compared to the surface of the Earth— well under 1%.

HB: In your *Very Short Introduction* book, you mention a long-standing debate in the marine science world that centres around the idea that on the one hand these are incredibly diverse ecosystems, and on the other hand, they seem to be quite fragile. So, the question is—

CS: Why?

HB: Yes. And one could spin this in different directions. One could say, "*Wow. This is such a diverse system that it should be sufficiently robust and be able to adapt to whatever's happening.*" Or you might say, "*No, it's so incredibly diverse that if you tweak it over here, it's going to affect so many other things, that it will be incredibly fragile.*"

What's your position on the situation? How would you assess it?

CS: My motivation is to simply look at what is happening.

If you think of all the reefs of the world, in rough proportions: about a quarter are dead; about a quarter are very badly damaged and at the present rate, are going to be dead in not too much time; a quarter are damaged a bit, but they would recover if they're allowed to, and only a quarter are left alive.

Now, we turn to the dead and damaged three-quarters, and ask, *What caused that?* We can look and see what impacts they've had. In many parts of the world, the impacts are very, very severe.

So, whatever the robustness of the system, it's been overwhelmed by the fact that you've had lots of kilometres of fish farm on top of them; or you've had dynamite fishing, or they've been heated over their tolerance, so they've died.

At the margins of that, though, you've had a certain amount of damage, and you've had a certain amount of impact. Not too much, but enough to start the mortality.

Now, nothing can survive being buried. I've had one example, which was early on in my career when I started to learn the cynicism of the political system: a reef had been buried under two metres of concrete. And I said, "*Well, it's been killed.*" My report was changed to, "*This reef has received some impact.*" That was my first exposure to the commercial consultancy world.

So, it's difficult to speak from the perspective of a coral how much of an impact is going to be very bad. Instead, let's look at what we've got. We've got a quarter dead, a quarter in good condition, and the rest sort of between the two.

Whether you call a reef "robust" or not, we are still impacting them more than they can survive. A lot of the time they will come back, if they can—if there are larvae there. Very often, in a lot of places we've seen, so many of the adults are removed, and the conditions are pretty hostile anyway because of overfishing, sewage, whatever, or a combination of all of them, that there are no larvae there. There are no juveniles. There isn't going to be another generation, so that's obviously beyond the tolerance of the corals.

And we know that corals have different amounts of tolerances: so you can knock out some species, and others might start to compensate a bit—I've seen that. But there comes a point when you dredge too much, you have more people and more sewage, and that kills the reef completely. It can't stand up to them.

So, whatever their tolerance, we're doing too much. They might be robust, they might be versatile—and they are both—but we can easily overwhelm however robust they are.

HB: What are the timescales involved? You mentioned the importance of the larvae and juveniles for the potential regeneration of a sick reef or a dying reef. What are the timescales involved, typically, for which there will be juveniles after some event has happened?

CS: We've measured this in the Chagos Archipelago. That is a reef not affected by human impacts. People were evicted or evacuated by the 1970s. Nevertheless something like 92% of them were killed in the warming of 1998.

We kept going back, and for a few years we just saw nothing. It was a heartbreaking scene because this was a glorious reef. We measured the recovery of the corals in terms of the cover of a reef formed by corals; and we found there were one or two what we call "weed species", which just grow maybe like bracken does, in a cleared area—it's exactly like that.

Now, these were branching, these formed glorious sort of tables. And by about 2006, 2007, they had got back to form this three-dimensional structure, which is so important to maintaining everything else. So the recovery back to "a functional reef", if you like, or what looked like a functional reef, took about eight years.

I don't think most corals will breed before they're about 2–5 years old. But there were obviously reservoirs around, maybe deeper, maybe in a cooler pass or embayment, where there were adults—enough to produce enough larvae to form the next generation.

So, it took about somewhere between six years and a decade for the reef to restore itself. Subsequently, it's been hit again by the next warming, which is just ending now; but that's another story, which is still a little bit unclear for some parts of the world, and distressingly clear for others where they have been set back to ground zero again. But on the whole it takes quite a few years.

HB: But, presumably, if we were to look at a reef that appeared to be effectively dead for 20–25 years, we would have some assurance that there would be no larvae that would be around there. Is that a fair assumption?

CS: That's right. There are areas of the world where the reefs are dead, and you do a larvae count and there are none there. Well, there might be the odd one, but not enough to make any difference. And you know that those reefs are gone. They're just not going to come back unless they can get a recruitment, an influx of larvae from somewhere else.

HB: Might it be possible to actually do that? Might it be possible, somehow, to reseed reefs?

CS: People have been talking about that. It's a matter of how realistic it is. You can have coral farms, and there are some around the world, to grow little bits of coral that you plant out.

In my view, that's fine outside the windows of an underwater restaurant, which I've seen, and it might be fine for some small areas. But you stand on the hilltop in the Seychelles and look out over thousands of square kilometres of reef, how are you going to do that? You might only need a few in the middle to then spawn everywhere else. Well, you might—we can keep our fingers crossed—sometimes I think that keeping our fingers crossed is the only form of management we've got. That's not a good way to run the planet, in my view.

HB: Let's talk a little bit more about the Chagos Archipelago. In terms of stewardship, this seems to me to be a very positive data point in so far as there was an established consensus in having a large-scale marine protected area. Can you tell me a little bit about the story there, and to what extent that might be a valuable precedent?

CS: Well, that archipelago is south of the Equator, and it's a line: if you draw a line from the Lakshadweep through the Maldives, Chagos is at the bottom of that same chain of a volcano in times past.

Now, there are five atolls with islands; and quite a lot more than five, which are submerged through natural reasons. It was made into British Indian Ocean territory in the late '60s, and the reason for doing that was the Americans created a military facility on one of those atolls—the one in the far south, in fact—which contains a lot of the landmass of the whole archipelago.

HB: Why did they do this, just out of curiosity—was this for nuclear testing back in the day?

CS: No, no. The idea was—and still is, I believe—to have a forward base, a forward supply base: it's a storage in the Indian Ocean that was created at a time when the Cold War was at its height.

The islands, all five atolls, had been planted by coconuts from the late 1700s, or early 1800s. The local vegetation was cleared, and they were planted throughout by coconuts—at least all the larger islands were. And those plantations did very well. The islands were called the Oil islands for a long time, up to the middle of the 20th century I believe. And then, because of mismanagement and other reasons, they started to fall apart. The cost was very high, the production was very low, and they went close to bankruptcy about three times, in fact. The first two atolls closed in 1935, the others carried on, but the population was going down.

Anyway, that archipelago was chosen to become a BIOT, British Indian Ocean Territory. And the base was constructed on Diego Garcia, the atoll in the south. It was built, to be fair to them, over coconut plantation—it wasn't as though the bird colonies were destroyed because of that, they had been destroyed when they cleared the local plants to build their plantations.

I first went there in 1975, and again for a long expedition, which lasted nine months in 1978-79; and it was a wonderful place to do research and to go diving. But at that time, the reefs there were certainly wonderful, but they were not too different to all the others we were seeing all around the world.

Then there came a period that some people have called the "Decades of Destruction", which affected everywhere where there were people. I mean, it's a sad indictment. Reefs then started to decline: too much fish were taken out. You take out the herbivores, you let the algae explode—they grow much faster than the corals, they want the same space, they'll out-compete the corals, to put it simply. And reefs started to decline around the tropical world.

The reefs of the Chagos Archipelago stayed in really good condition. There was no driving force to degrade them. And so the gap between Chagos and a lot of the rest really started to increase.

And that became very, very important the more towards the present day that we're talking about, because people are now increasingly asking, in Africa, in Sri Lanka, "*How do we repair our own reefs? We now recognize that they are essential for the production of food protein for countless numbers of people, but we've damaged them. How can we get them back?*"

So, we have here, now, an example of how reefs should look like, and how they should work. If you have a country where the budget for this sort of thing is very low—there's not much money in the country anyway—should the manager there control, or propose controlling overfishing, sewage, landfill and dredging? He could waste his entire budget doing something which isn't important simply because he doesn't know. You've got to see a real reef, a real unimpacted reef, to know what you're aiming for.

And that, increasingly, is being used to help guide restoration attempts in other parts of the world. There's nothing a local manager can do to control against warming, but at least he might be able to develop a deeper understanding: he might know that it's not the fishing, it's the sewage—or it's not the sewage, it's the fishing—or something like that.

HB: I'd like to explore the effects of global warming in more detail shortly, but before I do I think it's worth explicitly commenting on an important point about your tone, both in your writings and what you're saying now: you very deliberately stay away from a dichotomy between the environment and the economic necessities of those on the ground.

You consistently recognize that there are many people who depend very strongly on fishing, or other aspects of natural resources that the reefs provide. You are hardly advocating something along the lines of, *"Well, we have to protect the reefs, and too bad if people suffer as a result."*

Instead, you consistently emphasize that not only must we find a way to preserve these marine ecosystems, but by so doing we also ensure that the people who live there can thrive as best as they can in the long term.

Overfishing is a classic example of the perils of short-term thinking. Yes, everybody's hungry, and people need to eat, but if we eat all the fish immediately, and we don't prepare for the future, then we will naturally be faced with a very poor future.

In short, my sense is that you're very sensitive to the economic needs of the indigenous people, and those who live in the surrounding community, stressing that there is a way that we can actually safeguard and potentially repair marine ecosystems while fully taking into account the economic needs of local people.

CS: Yes. We have to hope; and I think we *can* do it: we could do it, it's more a matter of, *Will we?* The economic pressures on the reef are very high, and the need for food is very, very high. You talked about the "economic needs" of these people. It's not so much economics, as just the need for food.

But there are—I've personally witnessed this—there are strong economic forces to do industrial kind of fishing, which will have a harmful effect on the local people.

Most of the areas I work are in desperate straights, I am often asked to go to visit areas where there is a very bad need to restore, where people ask, *"How can we get this back? Why is the water coming through my living room?"* That is, literally, the sort of thing I am forced to deal with. The economic forces are often pretty ruthless; and I've been to field sites where you go through villages where people are dying, where they tell you, *"My baby died last week."* And it does bring it home.

Essentially, we know what to do to maximize the production—the surplus protein, if you like—which you can take from a reef. We know that. We know how not to kill a coral reef. We know that. It's a matter of convincing authorities, what they call the decision-makers, "important stakeholders"—to use that horrible phrase—to do something about it and to do what you say.

Time and time again, I said, "*You haven't got this source of protein for these villages, that's why the villages have been evacuated, because you've killed them by X, Y, or Z.*"

And the minister of the government might say something like, "*Oh, no one told me that was important.*" Well, we **have** till the cows come home, but we can't get through enough. There's this gap between informing the decision-maker, who might or might not be ruthless himself, he might be quite benign and have the best interest of his people at heart. But he hasn't had the information that people like myself have been shouting about for a long, long time.

So, the first thing we need to do, in my opinion, is to have a branch of the media to translate between the science and the people who make the decisions, because I don't run the world.

Economists, and politicians, and lawyers run the world, unfortunately. So, they don't know; and they come and go, they revolve. And we do the teaching, if you like, of these leaders, again and again, in the same place, again and again, on about a three-year, or five-year cycle.

And it's perfectly true, the new leader **didn't** know that to do this, that, or the other, would cause a downstream impact and remove the reef, which would remove the fish, which would remove the village. We've got to get a better connection between the science and the decision-makers.

It's no good saying that scientists can't get our point across. Probably, we can't, but we have a job to do. I think it's the job of the media to get it across. No one's asking the media to do the science research: we're just asking the media to do the media bit.

But too often we have to do it ourselves, and probably not terribly effectively—evidently not very effectively because the trajectory of any measure, any environmental measure you care to think of, is going downhill. It shouldn't be; and there's no reason for it.

It's easy enough to point to things like, say, greed by a multi-national industrial fishing concern, sure, or political corruption. That certainly happens, of course—I've certainly seen it.

But by and large there is hope to get across the message because people will do what they have to do to survive. It is no good going up to that person who is throwing DDT onto his area of sea to get fish and saying, *"That's naughty; you really shouldn't be doing that."*

HB: Right. He's trying to survive another week.

CS: He has no choice. The choice should have been by the politician, or the leader, earlier on: to conserve an area, maybe to create a "No-take area", so that fish can thrive in there, produce more fish, and export their larvae to everywhere else.

HB: Another often overlooked point, in my view, is the notion of nonlinearity. You mention in some of your writings about how a fish which is five times bigger than another has many more than five times the number of eggs than the other. Which means that we really have to make sure and take care of the large fish, because those are the ones that will be driving future success in terms of enabling people to survive in the future. And then by analogy, a flourishing reef is a very large fish indeed.

CS: Yes. I think that fisheries laws around the world, if they have laws, are upside-down in a lot of respects. The rules are: you throw back the little ones, and you can keep the big ones. And that is wrong. It's practical, but when a 10 kilo grouper will produce 2 million eggs a year, ten 1-kilo groupers will produce much, much fewer—orders of magnitude fewer. So, which one should you be protecting if you want the future generation of grouper to grow up to become available?

There is no way to make a fishing net, which will retain the big and only catch the small ones, so there's a practical issue here. The only way you can protect the big ones—which are the seed stock, the big fat mamas for next year—are to make a protected area where nothing can be caught. You'd have to protect the little ones as well, fine. But in that area, you would keep all of them, including the big ones, which produce the most larvae. So, we have a thing where our laws say one thing but the laws of Mother Nature say the reverse.

And where there's a clash between the laws made by us and laws of Mother Nature, then inevitably, Mother Nature will win. We can make whatever laws we like. We can say, *"Fish should breed more."* And some regulations around I have seen, are not much less fatuous than that, they really are.

The only way to ensure that we have protein to extract for people next year, and beyond is to have strictly protected areas. There might be other exotic ways where you farm the big fish in a tank, and seed the eggs back, but they know how to do it better than we do. Just let them grow. Leave them alone in some areas.

And you'll very soon see that such areas have a biomass of fish, including a sort of harvestable surplus that you can crop, which shoots upwards. In a space of about three to five years in the Philippines, for example, they've found several examples where it only takes that long before you get a greater income, a greater take of protein for the village.

HB: So, allow me to interject because I'm confused by this. Not by what you're saying—what you're saying makes complete sense to me—but by people's motivations.

I don't want to sound naive. I understand corporate greed and self-interested politicians. But what I don't get is why so many people, like those in the fishing industry, seem intent on creating policies that are so clearly against their long-term interests.

I mean, I'm guessing that those in the fisheries industry do not want to go out of business in five or ten years. Why is it that they are not loudly lobbying for the necessary measures that will protect their very livelihood? This I don't understand at all.

CS: Well, let me explain. I'm not going to mention any names, but essentially, some of them are multinationals. There are cases, which

I have seen, where you can fish an area out. And you might leave devastation in your wake, but then they will move somewhere else, or perhaps not even do fishing anymore but change their business into selling something else.

Let me give you one example of fish farms, shrimp farms in Southeast Asia. I've worked on this issue in Sri Lanka, in fact. To make shrimp ponds, you clear the mangroves and make a pond, and you grow the shrimp in that.

Now, that lasts for a few years. But after a few years, they don't work anymore because the soil around it becomes soaked with anti-biotics; and in some cases, you get harmful minerals leaching out of the soil there.

So, the corporation has to move on. And they leave behind it an area where there used to be mangroves, and then there used to be the shrimp pond, and now it's like a lunar landscape—and no more productive than one. But they will move on; and they'll move on until all the mangroves are gone. I'm not exaggerating. They will remove all the mangroves completely.

They will even, in some cases, pay villages to hand over their patty fields—and rice is a staple diet—to convert those into shrimp ponds, which will last a bit. When those are gone, they have to move on somewhere else. Or, being a multinational, they can move into other aspects of business.

It's the bottom line this year, this accounting year, that counts; you'll likely have an eye on the year after, and maybe the year after that, but you're being all wishy-washy if you think in the long term—it's more an immediacy thing.

And that, I think is the driving force behind the economics, which makes things so short-term oriented. And that is the problem. You're right: fisheries would do better in ten years' time if they took this or that course now. But they don't. They're thinking in much shorter time frames than that.

HB: And it doesn't seem that there's any sense of learning lessons either. I mean, this is depressing for all sorts of reasons because this message is basically the same that's happened over decades. So, I come from Canada, and it seems to be clearly acknowledged now that much of the fishing industry in the Eastern part of Canada, what they call "the Maritimes", essentially collapsed due to overfishing.

These once-robust industries became effectively decimated, due to a lack of appropriate management. If you go to these places you can see the direct effect on the people and their livelihood. Many of these people had been fishing for generations and generations, and now they really don't have any more fish. So, you would think that at some level, people would recognize, *Hey, wait a minute. We've seen this story before. We've seen it in local areas, we've seen it in regional areas; we've seen it in poor countries and rich countries.*

It's depressing to me that one has to be stating the same basic mantra over, and over, and over again: that if you don't look further than your own nose, you're going to be in big trouble in five or ten years.

CS: There are no excuses for that, but it does happen again and again. And in the tropics, it's easier to do, I think, than in the developed world where we have policemen. It goes back even further than what you're saying. In Britain, and in France, in medieval times, we hanged people for dredging the seabed. Now, we subsidize it. We know exactly what dredging is doing. So, why do we still do it? There is no answer for that, except, I think, to get the people just to say, *"Enough is enough. I'm not going to vote you in."*

But you see, a lot of these places don't have a vote. In one country—I won't name it—I was talking to the Minister for Fisheries, and I asked to see the figures. They were brought in, and they were written on the back of an envelope. Now, you've heard the phrase, *Oh, you could do that on the back of an envelope?* Well, these really were. Often the information isn't available to them. Sometimes, too, a company will simply pay for the licence to what it wants to do, and it's legal. They've got the licence.

There is no excuse for a great deal of this, I quite agree. But, as I said, it's been going back much longer than things like the cod fisheries that you alluded to. And it's terribly depressing to see that happening still.

HB: You had mentioned the media earlier, and I think it's perhaps time to talk a little bit more about that because it's a primary vehicle by which people can become informed, and in turn hopefully put some pressure upon their political masters, certainly in the developed world, but also in the developing world to some extent. It seems fairly clear to me that you don't believe they're doing, in general, a terribly good job of that.

CS: That's correct. I'll tell you why they tell me they're doing such a bad job.

They would say that the public is not interested, that they're much more interested in what the local footballer, or singer, has done, or some embarrassing escapade a politician got up to. They'd say, *"Well, we have to respond to our audience."*

HB: Is that a justifiable response on their part, you think?

CS: I don't think it's justifiable, no. There are some media outlets, which not only have the responsibility, but were created to inform, such as the BBC. The amount of time that is given to something which is crashingly important is trivial compared to other things which grab the attention.

There are some issues, such as large influxes of refugees caused by civil wars and so forth, which are very important and quite justified in getting the attention they do. But there are also an awful lot of non-news items, to my way of thinking.

It's also true that people don't like bad news. Several journalists have told me, *"My readers don't want to hear about that. Can't you dress it up in a rosy way?"* Well, in many cases, no—there is no rosy way. In Malaysia, the fishing villages had been abandoned. There was no rosy story to that at all. And then in other countries journalists aren't even allowed to talk about such things, since they may reflect badly on the government. That happens quite a lot.

HB: OK, but let's take that out of the equation for the moment, because that's a far thornier issue. Let's just look at countries with a free media, who for whatever reason are electing not to do as good a job covering these issues as they could be doing—not only the current state of affairs, but also what can be done about them.

I'm guessing, reinforced by what you've said, that one concern might be people saying, "*That's all terribly depressing. I'm going to turn the channel because there's nothing I can do.*"

CS: Yes, I think that's a big factor.

HB: Another theme that you sometimes hear is, *Well, it's all very complicated. It's all very difficult. There are so many different factors that are involved. All this environmental degradation is terrible, of course, but of course, people have to eat.*

They make it seem as if it's some incredibly complex system where if you were to invoke some appropriate environmental regulations then some non-trivial proportion of the civilian population would become destitute and die. And so, you get this sense of, "*Oh, it's just a big mess and we can't even make head or tail of what the way out is.*"

CS: That's right. Several times I've seen a deliberate obfuscation of the issue. Very often, the science is quite easy, in terms of understanding, in a particular case, in a particular area, what's got to be done to stop things from getting worse. It might be quite simple and straightforward. The science might not be the problem at all: anyone can understand it, even the local politician.

But they might choose not to take that course. They might say, as you said, "*The jobs are going to depend on this factory. Okay, it's going to kill that estuary, but that won't be for a while.*"

I remember talking to one senior person about erosion of their shoreline. Behind that shoreline was a tank farm, an oil farm. And I was asked, "*How long would it be before the tank farm got undermined?*"

Well, I hate that kind of question because you're afraid, of course, to put a precise number to it. But I said, "*Probably about a decade or so."* And this particular leader responded, *"Oh, well that's alright then. I'm only in office for another five years."*

And you can be taken aback, or you can expect it. That was a society where I might have expected that sort of answer.

HB: It's honest, at least. You have to give him that.

CS: Well, yes, I suppose—full marks for honesty there: it'll be someone else's problem.

But to go back to your question regarding the media: often, the science is quite straightforward, but I don't underestimate the complexities of translating that in ways which would be effective, acceptable, and create action by the leaders for their people.

And very often, there might be some people for whom it's too late: they have to move, they have to go. The mortality now attributed to climate change is a nebulous figure, but there are more estimates all the time. They can be in the millions per year.

In some parts of the world there is famine. Now, if you do a google search, you can see a very depressing table that itemizes the Bengal famine, the Ethiopian famine and so on. They all have a start date and a stop date and the numbers of people estimated to have died.

The trouble with the one that we're dealing with here, in the over-fished, over-exploited parts of the world, is that there's no clear start date and certainly no clear stop date. It's continuous. But the numbers of people estimated to be dying from resource shortages caused by some of these impacts that we've talked about can be millions a year.

So, I would say it's a famine. People say that Malthus was wrong. Well, he was in so many ways, but you won't see this sort of thing written now in that Wikipedia table of "Famines". There's a sense that, *They're just dying because that's the way it's always been.*

Things are getting better in some places. But in some parts of the world—too many—the numbers of people dying from resource shortages is increasing. It's not going down. Some economists try to get a handle on this as well. I talked earlier about *eco*—ecology and economy—and how they should be interwoven much more. Were they ever? I don't know.

Questions for Discussion:

1. Do you agree that the media is, on the whole, doing a poor job of communicating environmental issues? If so, why do you think that is?

2. Should scientists such as Charles have a greater media presence than they do? If so, how should they go about doing so?

3. Do you think that some forms of government are more suited to developing and implementing long-term environmental solutions than others?

IV. Towards Progress?

Leadership, policies and philanthropic foundations

HB: Are there any positive signs? Economists who are looking more closely and productively at environmental factors and positively influencing public policy by having the ear of presidents and prime ministers? Do you have any sense of optimism that we might be somehow, somewhere, closing that gap between ecological issues and economic issues?

CS: Yes. I've seen a lot of examples around the world where the right things are being done. I would say that of all the sort of consultancies I've done, probably, I win about half the arguments—only half, but half is pleasing.

There is a caution there, that a win in environmental terms is only a win *for now*. The next dictator might decide to obliterate the place through negligence, not understanding, or something like that.

But I think that there is undoubtedly an increase in general awareness amongst the people who can make a difference, or who can influence things; there is definitely an improvement.

So, we can see it. It can be done. And it *is* being done by armies of us in the aid business, if you like. How can we increase that percentage? That's the question.

And linked to that is: *We know the answers, we know what you* **should** *do Mr. President, or Mr. Minister, but* **will** *you?* And there are other issues he considers—wealth, jobs, a whole range of things, or his own interests—which might mean that you know jolly well it's not going to be done there. But you've passed on the information.

I can think of two or three countries where scientists have come up to me and have said, "*We wish that you would start to colonize us again. It's our only hope."* So, I have to say, *"Well, I don't think that's going to happen."*

But that's a clear sign of how desperate some people are to say such things: they believe that their own rulers are so uninterested in their status that they know nothing will happen without a regime change. Now, that's a whole can of worms.

HB: Yes, but it does bring up, at least obliquely, another angle, which is the impact that global government and international organizations might have. Is there a constructive role for them in all of this?

CS: Yes.

HB: What, exactly, and how?

CS: The way things are increasingly done is to provide "a development package" so that you can increase the "capacity", as it's called: the scientific ability, and the status of the local institution—the local scientists and things like that. The idea is that those like the person who said to me, *"Our only hope is if you would colonize us again,"* would become more influential and more respected, not just at the bottom or close to the bottom to the heap with those subsistence livers—well, they're about halfway up, actually, the academics.

HB: Although, they're still getting less than a dynamite fisherman.

CS: Yes, well, that determines where the clout lies, you see. In a case like that the fisherman could pay the police station, or the police officers in it, to turn a blind eye to what he's doing, even though it is actually illegal.

So, I think we need cooperation in the form of an empowerment, and improvement of that local research institute, in terms of both the research it is doing and the impact of the advice it is able to give to the senior echelons.

That is where the influence comes because I don't think a lot of this is a scientific problem anymore. It is a problem of politics and sociology. That's where the problem lies.

We know, so often, what the answers are. Very often they're unpalatable, you see, to some sectors or another. And if they're unpalatable to the sector which has the clout and does the determining, then they will be ignored.

Somehow the whole status of the middle people in much of the world, who know what is happening every bit as much as I do—more, in fact, because they see it daily—*that* needs to be improved; they need to be listened to much more.

HB: Is there also a role for global leadership by some of the major powers? Earlier you mentioned the British Indian Ocean Territory and before we started filming you spoke about some of the recent efforts of the American government. How large a role can an American President or French President or Japanese Prime Minister or coalition of Western Leaders play in impacting policy?

CS: I think they can play and are playing a very big role. They're already in some of those giant protected regions we spoke about earlier, but it's still got to translate down to the area run by a local village.

But on the global kind of scale, some of the Western states *are* setting very good, very necessary examples. We can afford to set aside these blue belts around some of our overseas territories, and Hawaii, and others as well. They will benefit *everyone* in the end because they're big enough.

There have been program, for example, which identify very large areas should be protected. The criteria for doing that is they've got to be big enough to make a difference if they were protected, they've got to have a government that will enable it to be protected—you can declare what you like on the outskirts of major city in the tropics, but it's not going to help it—and it's got to be in good condition to start with. It's got to be worth protecting.

And the very big marine protected areas are a response to that need. And they are happening. And with export—species flow and things like that—the idea is that will help everyone in a much wider area of ocean than they are located in.

It's just like any protected area. You protect the fish there and the export of larvae outside it will be effective, it will make a difference. I referred earlier to examples in Southeast Asia, where after totally protecting "their reef", then a village will see after a few years, an increase in both income and protein. So, we know they work.

My main goal would be to see a marine spatial planning system, which includes areas that have already been sacrificed—completely covered with fish farms, say—through to multi-use and right up to "no take" protected areas: the whole suite, rather like Western countries do with agriculture now. You have your national parks; you have an area which *is* a farm, with almost no diversity to speak of other than corn. And you'll have some areas where you can't do anything. I think that's the type of solution we've got to translate from land to sea.

It's already happening in a very piecemeal way, and a lot of multi-national aid agencies know this, they're working towards it. You encroach on issues of sovereignty and a lot of things if you do that.

There's opposition to a lot of things. A lot of environmental opposition happens these days, from the rather dotty climate change denial business, which is now generally regarded as a bit like aliens and flat Earth.

But there's also opposition to, for example, the Chagos Marine Protected Area. The opposition there comes from a number of sources. Very often, some of the most vitriolic that I have faced—because I was one of the people working in the background to help create it—has come from people who just hate the idea of the American military, or the Western military. So, in their view anything done there is going to be wrong to start with. Some people have told me that outright: "*Well, nothing you do is going to be right.*" And I'm associated with "a capitalist, imperialist machine".

HB: They have a one-issue agenda.

CS: That's exactly right. And the environment is going to suffer. We know that this archipelago is very important in the Indian Ocean as a whole. It's right in the middle: it's a stepping stone of the East-West flow. Half the year the current flows one way, half the year it flows the other way. It's crucial to a much greater proportion of people than many appreciate.

And that applies to a lot of marine protected areas. So, I'm a strong advocate of those as one of the remedies because you can see what good it does—not just for the science, for understanding that place—but also, on a more local level, how much it can actually benefit other areas which are downstream.

HB: As you were talking, I began thinking about the role of philanthropic organizations in all of this. I understand that there is a particular foundation, the Bertarelli Foundation, that has funded some efforts related to what you were mentioning?

CS: Yes, the Bertarelli Foundation funded the patrol ship for the BIOT area; and they have also funded, and are funding, and will continue, I hope, to fund research there. Their focus is perhaps a little more on the ocean fish, the big fish, the iconic species, sentinel species, rather than on the reefs and islands themselves. That's their focus. And there are other foundations that are funding research as well.

HB: They strike me as representing a happy medium, or a happy connection at any rate, between government, research scientists and like-minded people who are concerned about marine stewardship, and the stewardship of other environmental ecosystems.

CS: I think that's right. We see it in many parts of the world. Let's face it, a lot of the world's countries just don't have any money. If you listen to our government, we don't either. And the philanthropic institutions that fund detailed programs of research are incredibly valuable. They represent a very important adjunct to what the government does; and in some parts of the world I can see that the philanthropic side of it is greater than the governmental side of it. Perhaps in some places, it always has been. I'm not sure. But certainly, it is a very significant amount of funding.

In any given country there will be an appreciable number of NGOs who are doing useful and important things there—not just the urgent aid things like feeding starving people in a famine, but also what we call development aid, which is to try to foresee what's going to happen next year, or the year after: getting measures in place so that a crisis *doesn't* happen. These are funded both by thousands of people who regularly donate resources, putting their money in the tin, as it were, as well as the really big, well-endowed foundations.

They are both incredibly valuable, and they fill a need that governments, even wealthy Western governments, seem not to be able, or not willing, to do.

Questions for Discussion:

1. Are there some issues that NGOs are better placed to tackle than local or regional governments? Worse placed?

2. To what extent do you think the environmental movement is particularly prone to being too inflexible and not pragmatic enough to meet pressing objectives?

V. Climate Change
Two pernicious effects

HB: And then there are the issues such as global warming, which are, of course, well beyond the scope of any one particular region, or government. Maybe now would be a good time for you to delineate the specific adverse impacts to marine ecosystems caused by climate change, as opposed to other factors that can be regulated and controlled, at least in principle, such as fishing.

CS: Sure. Let's take warming first. As we form the CO_2 blanket, as that increases, warming increases. We can look at the CO_2 graphs, and they're going up, and up, and up. Now, warming puts a lot of animals and plants over their survivable threshold. Corals are one such animal. Their temperature depends on the sea temperature. They don't control their own temperature. The warmer they are, the faster they work. They work at the top-end of what they're able to, in terms of their metabolism. When you get a warming pulse, that puts the temperature over their ability to live.

HB: So, how high is that? What's the upper limit for them, more or less?

CS: It varies around the world. But let's take the Chagos Archipelago. The normal temperature up to about a decade ago was never warmer, probably, than about 27° or 28°—in a lagoon, it might get to 29° or so. Every species will have its own exact, precise threshold.

Now, it's warmed a degree or two more than that over periods throughout the warm season in that area, and that is enough to kill them. So, it only needs a degree or two above what it should be to have killed the corals, and it has done that—more than once. So, basically, it is warming that is causing those problems. That is an immediate effect—a short-term effect, years to decades—and that is happening.

The other thing that happens with a build up of CO_2 is that more of it dissolves in the water. It's the same system, if you like, as in our blood: you dissolve the CO_2, you form a carbonic acid, that adjusts itself into bicarbonate, adjusts itself to carbonate, and the carbonate is what makes the reef because the coral animals take that out of the water to deposit it.

When you increase the acidity—that's a bad term, actually: what is happening is a decrease in alkalinity. It's still the alkaline side of neutral.

HB: You're relatively increasing the acidity.

CS: Yes. You're relatively increasing the acidity.

And when you do that, things like corals—and a lot of other things as well, incidentally, like plankton groups, which are crucially important to the oceans and engaged in similar activities—have to work harder, use more energy, to get the limestone out of the water, and make their own shells.

And the acidification is such that when you reach a certain point, they just can't do it anymore. There are several forms of calcium carbonate. The ones that the corals, and a lot of the plankton, use is called aragonite, which is a certain crystalline form of calcium carbonate that's the most sensitive to acidification. And there doesn't have to be much of it to result in the organism ceasing to be able to deposit the limestone.

HB: And presumably, these factors are not only very dangerous individually, but the combination of the two can result in a particularly malevolent threat to these organisms.

CS: Exactly. It's like the diseases I talked about before. If you have the two together, that's more than doubly bad—it's often a synergistic effect.

So, those are the two global issues.

The warming is affecting the tropical oceans quite significantly because the species that live there are already close to their thermal higher limit; and it doesn't need much, just a degree or two, to push them over it, and then they fall over.

Acidification is such that—this a chemistry thing—CO_2 will dissolve in colder water more easily than it does in warm water. So that is affecting the poles more. Acidification in the poles is thus probably happening faster than in the tropics.

And the two give a sort of nutcracker effect: you have the warming impacts greater at the tropics and the acidification factors greater at the poles, and the two are both spreading in range.

Questions for Discussions:

1. Were you aware of these two effects of climate change on corals before you read this chapter? Given how straightforwardly intelligible Charles' description is, are you surprised at how little known it is?

2. What does Charles mean, exactly, when he talks about "a synergistic effect"?

VI. What To Do?

The importance of marine spatial planning

HB: So, I can imagine that somebody reading this might feel very depressed and indignant at this point, convinced that the situation is hopeless and there's nothing that can be done. What would you recommend in terms of concrete action?

CS: A bit more activism. A lot of people might be indignant and believe that there's nothing they can do. But a bit more activism, I think, is essential.

In many of the countries I've worked in, the scientists have told me that's what they need: they need pressure to be put on their own systems. There are lots of reasons why you could say that such a thing can't happen, or probably won't happen, or might not happen—but the truth is that some of these places need it. And they also want it.

Now, the government there might not want it. So, it *is* depressing, but again you can use a medical analogy: I know one day I'll be dead, but it won't stop me going to see the doctor tomorrow if I need to. Why do I do it when I know I'll be dead in 50 years time? It's pretty pointless, isn't it?

Well, I think the analogy fits here as well. You keep doing it, you keep trying. And very often, as I said earlier, about 50% of the time, I manage to achieve the desired result in a consultancy for an oil company, or an EU or UN organization. You just keep doing it, don't you?

HB: What sort of concrete advice would you give to somebody who's not in the field—our neighbour, say, the person living next door to us in North London. What should that person be doing? Putting pressure on her local MP? Writing letters to the editor of her news-paper? Engaging in crowd-funding initiatives on the internet to support environmental activities? What sorts of things would you recommend?

CS. I know many people who do all those things, actually. I think that because the indicators are still not looking upwards, there are not enough people doing that. But, yes, in a lot of cases, that is what should be done.

People talk about education. That's crucial. I've been in the education business for a long time at university. That is core to a great deal of it. But, you know, I was thinking the other month at a lecture I was doing, we've been saying that for 40 years, that I know of. We're not doing it enough, basically.

HB: Are we doing it better? Are we doing a better job teaching young people about environmental awareness and environmental issues than we did before?

CS: I'm not sure if the teaching comes through schools. They have a curriculum, which is pretty straight-jacketed—in this country anyway. My wife's a teacher and she has told me that. That you can propose extra-curricular things for school, and the teachers will come back and say, *"Well, we can't even fit in what we're legally obliged to do, so we can't. Thank you very much. It might be interesting, but we can't do it."*

There has to be a change in the awareness that it's important to **us**. The remark I made about this particular minister saying, *"They're a long way southeast of Dover"*—that has got to be overcome. And it does need a lot of people to write to them. I once heard that it doesn't need many letters to a member of Parliament, for him to think it's a big issue and do it. And if more people did, I think they would do more.

Britain does do pretty well in a lot of areas in this respect. We now are undergoing convulsions about protecting areas around our own coastline. There's opposition, it's been diluted, and all sorts of silly things are going on, which are quite inexcusable considering we should know better.

But in contrast to that, Britain is creating these very large protected areas around some of the overseas territories. And that is a very positive step.

But they say now that it takes four barrels of oil to catch one barrel of fish in the North Sea. It's very over-fished; and that's because it's subsidized. As I said earlier, they used to hang people for trawling, now we subsidize it.

It's just *got* to be recognized that we could get back if we let nature do it itself, and we used restraint, now, on the massive kind of catch that we're taking out of the sea. In some cases around the world, it may well be too late. But let's learn from that, for heaven's sake.

Here, we should as well. We have something called the shifting baseline syndrome, I don't know if you've come across that. But if a change happens gradually, you don't realize there's been a change. I've seen environmental assessments where people go out to a place, which is pretty badly damaged, and do a baseline survey to see what it's like, what they say it should be like, and any change is measured from that. It shouldn't be. It should be measured from a baseline survey, if it was done, 50 years earlier, a century earlier.

HB: When it was actually healthy.

CS: There has been change. You've all heard older people saying things like, "*In my day, I was catching fish three times bigger than what you'd get today.*" Now, is this just an old man grumbling? Well, it might have been, but probably not. Nowadays, at any rate, we've got the numbers to show it.

I've been on a reef where it was just a seaweed plinth, really. And people were there from Britain who were doing what they called "a baseline survey", and they were ecstatic. They'd only ever dived in the North Sea, or something, before. And they could see a fish. It was warm, they weren't shivering. It was wonderful, you know? Well, that's a shifting baseline syndrome.

It shouldn't have been like that, it should have been a glorious reef like we could show them all over the place in the Northern Red Sea. But if that assessment was used as being what it *should* be like and any change of the factory was going to be measured from how much it goes on from that, they are on a losing trajectory.

HB: Indeed. Let me try now to be as concrete as possible about what should be done. Earlier you lamented—completely justifiably, in my view—that the world is run by lawyers, and politicians, and economists. But for the next few minutes, I'm giving you the opportunity to redress the balance: I'm putting you in charge of the world, as it actually is now with all of its present environmental degradation, but giving you the power to design and implement whatever legislation or agreement you wish. What would you do?

CS: I would develop, to a very great extent, the whole concept of marine spatial planning. Much of the problem of the oceans, that the oceans face, is because of a lack of regulation. Don't forget, most of the oceans is not in any national jurisdiction for a start.

So, with marine spatial planning you will have areas, like on land, which you have set aside to act as a larder for recruitment, to let the big fat codfish or grouper grow up, and produce all those extra millions of eggs. You would have areas that would be zoned in all sorts of complicated ways, including areas where you would say, "*Okay. That area, that whole bay is going to be 1,000 square kilometres of fish farm.*"

It's kind of happening like that in an ad hoc way, but I would do it a regulated, planned way, in a much sensible way.

Right now, we're forced into sacrificing an area because it already has gone. We're forced into making a marine park in some area or another because you *can*, not necessarily for the best reasons, but because it'll be achievable. So, the most important thing is to get that organized, and accept the fact that if we go piecemeal at it, as is happening in most of the world now, then all we will get is a net degradation.

If we can plan it, like with farmland—a national park here, a farm there—we can ensure that we will both retain that national park and the agriculture. We need to do that for the ocean. I think that's the key to it. It'll happen in an ad hoc, very sub-optimal way anyway. It'll be forced on us, especially with the rise of human population. So, let's organize it and plan it.

The problem here is that a lot of the ocean, as I said earlier, is not under a national jurisdiction: it's very often under the de facto jurisdiction of the factory fishing fleets, who will discharge their big hauls of fish in states which might be less than wonderfully adhering to global norms in what they should and shouldn't be receiving.

The point is not so much to get into the fisheries issue, but rather to highlight the fact that there is an absence of jurisdiction, a lack of any authoritative body to be able to say, "*You can do this, or you can't do that.*" So, the response by others will invariably be something like, "*Well, no one's telling me not to.*"

The other thing I would do is reverse the burden of proof. Right now, you can often cause damage, unless you can prove it'll do harm. I'd like to see that reversed to, "*No, you can't do it, unless you prove it's not going to do any harm.*"

HB: And this speaks to what you said earlier about the fact that in most cases, we know very well what's going on.

CS: We do.

HB: We have a clear scientific understanding of causes and effects.

CS: I've been in many situations when I've heard a company say, "*Well, there's no proof it'll do harm,*" and I replied, "*Well, up the coast there, we know that a similar thing **has** done harm.*" And they'll just say, "*No, that's not proof it's going to happen here.*"

That's completely fatuous. We **know** what will happen if they do that. The burden of proof must be turned around, saying that the mechanisms they're going to use to do their development **won't** cause harm.

Now, I should say that this development in question might bring benefit to huge numbers of people. I'm not anti-development—people often make that easy mistake. But there are better and worse ways to do things, and the better way might sometimes be more expensive than what the company had in mind at first.

So, we must change the burden of proof around: *"I'll let you do that if you can show you're not going to do harm, and you've taken all the measures needed to ensure that's the case."* Because the Chief of the World has the global spreadsheet. The company might find it cheaper to do it, let's say, "the bad way".

And the extra costs are externalized, aren't they? So, the villages who are displaced, that's an externalized cost. The company's not paying the bill of the consequences of that, but the ruler of the planet, or even of the country, would know that it's going to cost him or her more if the village is displaced and they have to do something else somewhere else.

It might not be company's bill, but it'll be their own social structure bill. It might not even be dollars you're talking about. It might some unmeasurable, unquantifiable thing. How do you price the value of a human life? That sort of thing.

People have, incidentally, priced values of human lives in lots of different places—it's pretty grim—but even leaving that aside, if you have control of the spreadsheet of the world, you know that doing it the bad way that is cheaper for the company will be many times more expensive for the ruler of the world.

HB: Right. Is there anything that I've neglected to mention, or we haven't spoken about enough? Anything you'd like to talk more about?

CS: Well, I would comment, perhaps, that people going into the line of work I've been in—whether it's land or sea—need to know the science, but they also need to know quite a lot, as much as they can, about how systems work in the countries they're going to.

It's very easy for a consultant from the West to breeze into a place, saying, *"Do this, do that,"* and then fly out again. That doesn't work, it doesn't help.

They've also got to be robust in themselves, because the vested interests can be fairly powerful and very strong, and you get a lot of ad hominem attacks. Often, they backfire in the end. In fact, in a recent Supreme Court, in Britain, judgement, my alleged "incompetence" was one of the planks they used to say this, that, or the other; and I was delighted with the judgement that came down, which was quite the reverse. But you've got to be robust because people will try whatever they think might work: if they can't rubbish the science, they'll try to rubbish the scientist. And that happens a lot.

A lot of people I know have been put off completely by doing that. But hey, you keep on.

HB: And it's vitally important, of course, for successive generations to be sufficiently stalwart that they don't get dissuaded from going into this extremely important field because of the ad hominem attacks that will be waiting for them—and those attacks will likely be ever sharper the better you do your job.

CS: Yes, that's my point. There are a lot of obstacles that will be put in your way. Well, sometimes there aren't, sometimes everyone's on the same side, they want to fix something. But very often, some of the problems are caused because someone has caused them; and you want to arrest the damaging activities of that someone.

Now, that someone, if there's significant environmental damage that's been caused, is a significant person in that community. And my point exactly, is that the science is one thing—it's an essential grounding—but there's a lot more besides. But I would say, "*Keep on doing it. It's very much worth doing, and good fun in the end.*"

HB: Well, you seem to have weathered the storm remarkably well, I must say. It's been a great pleasure talking to you, Charles—I've kept you here for well over two hours now?

CS: Was it really? Gosh. I enjoyed it too.

Questions for Discussion:

1. Should the world's nations formally designate an international body with jurisdiction over the world's oceans? If so, what do you think are the obstacles to such a thing happening?

2. How might we, as a society, better encourage people to be involved in the sorts of activities that Charles is engaged in?

Continuing the Conversation

Readers interested in more details on coral reefs are encouraged to read Charles' related *Very Short Introduction*, published by Oxford University Press as well as his co-authored book, *The Biology of Coral Reefs*.

Saving the World at Business School

Part 1

A first conversation with Andy Hoffman

Introduction

Clarity vs Popularity

A friend of mine has long argued that there is an inverse relationship between the popularity of a word and its meaning. The trendier a word has become, he says, the fuzzier it is, until eventually it's used everywhere and means nothing.

"Sustainability" seems a perfect example for his theory. Once a word primarily associated with dour environmentalists, it's hard to think of someone these days who does not avidly chatter away about its merits. Politicians of all stripes routinely vie to outdo one another to demonstrate their sustainability credentials. Corporations now have Chief Sustainability Officers. We are all sustainability advocates now, it would appear. But what, in fact, are we actually talking about?

Not much, in fact.

Into this yawning semantic void steps Andy Hoffman. A popular professor at one of America's elite business schools, Hoffman might seem an odd choice to be the driving force for a fundamental re-interpretation of the green lexicon.

But a closer examination shows that he's spent the majority of his career searching for constructive and practical ways to develop mutually beneficial common ground between the forces of capitalism and environmentalism. He is, after all, the Professor of Sustainable Enterprise at the University of Michigan.

There's that word again.

But Andy, to his credit, keeps pushing our understanding of what it actually means.

However vague it might be, he told me, our widespread invocation of "sustainability" is clearly a good thing. Once rejected from a position at a top-tier business school for being "too focused on the environment", he has witnessed first-hand the evolution of the environmental movement as "sustainability" has moved into the mainstream.

But for Andy, that journey is only just beginning.

> *"Now it's time to discover 'Sustainability 2.0'. Where do we have to go next? There's been change to a certain point. But the problems continue to get worse and even more radical shifts are called for."*

A radical shift is exactly what you might call the recent work he co-authored with his mentor John Ehrenfeld, *Flourishing: A Frank Conversation About Sustainability*.

The book is a dialogue between the two experts, beginning with an analysis of the issues at play and concluding with a final chapter, *Reasons to be Hopeful*.

Throughout the conversation, Hoffman plays the straight man to Ehrenfeld's more radical declarations. What is needed, Ehrenfeld avers, is not simply incremental improvements to help us preserve our status quo, but nothing less than a redefinition of our core values, a collective societal shift away from perpetual consumerism towards a deeper understanding of our place in the world.

To that end, a new definition of that oh-so-troubling word is presented. "Sustainability", we are told, "is the possibility that humans and other life will flourish on the Earth forever."

As each aspect of this quasi-utopian announcement is examined, scrutinized and dissected in the cold light of day, I felt myself increasingly at sea. Being confronted by an unabashedly idealistic tract that boldly announces a clear road map for societal progress is one thing—we all need to be inspired from time to time.

But what on earth is the world coming to when these sorts of things are being written by two *engineers*, one of whom is a faculty member at a major American *business* school? What's next? Disarmament pamphlets from the NRA? Vegetarian cookbooks by the French? Scandinavian samba videos?

The truth is that I've never had a clear understanding of what happens inside business schools anyway. I knew that fees were high, and that their shiny, modern buildings were often populated with people with PhDs wearing suits, which has always struck me as vaguely oxymoronic. And, in stark contrast to the atmosphere pervading physics or philosophy departments, say, most business students seemed convinced that time spent there might well lead to an actual job.

That was about it.

And then there was the fact, of course, that all business students were superficial, morally-depraved, mindlessly-consuming sell-outs who were largely responsible for driving the planet to the brink of ecological destruction.

But this last point, I discovered when I had the chance to sit down and chat with Andy, needed a bit of a rethink.

> *"Well, there's definitely a demographic you describe, but more and more students are coming into business schools because they want to make a positive change in the world and they see that business has the power base to do it. They see the potential opportunities.*

> *"When I first got into this, I wanted to try to teach students to go into companies and help them to see environmental issues as strategic opportunities. Now we have more and more students coming out and saying, 'I don't want to go into a company and teach them, I want to do it myself. Increasingly, young people are motivated by the idea of creating a company that can try to address social and environmental issues.*

> *"There's also a focus now on the 'hybrid organization', the sort of organization that lives in the blurry space between the for-profit and nonprofit world. We're seeing more and more students who want to*

do that. They want to make a positive impact on the world and they see a business as a way to do it."

All very smoothly delivered—he is a business professor, after all. But Andy Hoffman is clearly no ordinary guy in a suit. Before returning to do his PhD at MIT, he took 5 years off to become a carpenter and home builder, a story he detailed in his award-winning memoir *Builder's Apprentice.*

And while it's worth emphasizing that virtually all of the truly radical things offered up in *Flourishing* clearly spring from Ehrenfeld rather than his erstwhile student, it's equally obvious that Andy's strong resonance with his mentor's views was a prime motivating force for the book's creation in the first place.

"I see John as a visionary. I see him as looking much further out than most of us can see. He's a very deep thinker, he's been thinking about these issues for a long time and he's very philosophical.

*"What he's pointing out is where we need to go for the long term, that some things we're focusing on now are not going to take us where we need to go. Yes, you can buy a compact fluorescent light bulb and screw it in. That's great, you're reducing your energy load. But there are still a lot of materials that went into that: you're just making the production of light **less bad**. How do we shift from there to actually making our technological society **better**?*

*"He makes the really powerful point that all our efforts right now are **reducing unsustainability**, which is a fundamentally different thing than **creating sustainability**. He's not against windmills and compact fluorescents and hybrid cars, because he says that's slowing the velocity at which we're heading towards that brick wall. But if we want to stop and reverse course, we have to think fundamentally differently about this problem."*

From a rather unexpected quarter, Andy Hoffman offers us a penetrating diagnosis of one of the biggest problems of our age. It might well not be a terribly popular one at the moment. But perhaps that's just evidence of its inherent meaningfulness.

The Conversation

I. Building a Career

Getting a lay of the land

HB: I'd like to start with your background, how you got into your present situation. Your undergraduate degree was in chemical engineering, right? And then you moved to civil engineering and management, I believe.

Throughout all of that, you seem to have maintained an interest in both environmental issues and the corporate world. How did that all start for you?

AH: Well, the environmental awareness started when I was an undergraduate. I had decided to pursue a chemical engineering degree without too much reflection: I liked chemistry, I liked math. I put as much thought into that as any 18 year old would.

And then Love Canal happened when I was an undergraduate. I thought to myself, *That's something I can use my chemical engineering training for, to make sure that sort of thing never happens again.*

So I minored in environmental engineering, which at that time was just waste-water engineering. It wasn't focused on pollution, or anything like that.

My first job was with the EPA (The Environmental Protection Agency), and I hated it. I worked there for two years and just felt like I was making paper. A funny tangent to the story is that I decided that I needed to be higher in government to have an impact. So I applied to Harvard and Berkeley for public policy, got in, and then froze. I couldn't get myself to do it.

I helped a friend build a deck at the time and got a charge out of it, so I started scanning the classified ads in the newspapers and eventually got a job as a carpenter in Nantucket. That turned out to be Jack Welch's house, the CEO of General Electric, and within 2 years I was supervising a 29,000 square foot house in southwestern Connecticut.

I did that for 5 years and then decided to go back to graduate school for construction management. Environmental issues got exciting then, since businesses started doing it because they wanted to. When I was working at the EPA, I was just a policeman: it was just a pain when I showed up and ruined people's days. But now it was strategic. I was offered the chance to do a PhD at MIT and took it.

HB: And when you were a builder for all these years, were you still reading passionately about the environment? What was your mindset at the time?

AH: I was going in a totally different direction. Occasionally, there would be overlaps with environmental issues. I was once a permit writer for a facility in southwestern Connecticut, which turned into a Superfund site. I knew who the local activists were and I called them up and said, *"I'm Andy Hoffman, and believe it or not I don't work for the EPA anymore."* They were suspicious. They said, *"Let's meet with you first."* They had decided I was a plant for the company and they wouldn't meet with me.

That was about the extent of it. I was really devoted to building and it wasn't terribly environmental. I mean, 29,000 square feet? No one really needs that kind of space.

HB: So you went back and did your PhD at MIT in Civil Engineering and Management, where you were able to get some of those old fires rekindled, presumably. How did that happen?

AH: Well, it was just the idea of trying to focus on positive change, rather than mere negative enforcement. When companies started to see that there was a connection between their strategy and their ultimate interest in protecting the environment, that's when it got really exciting.

There was a lot of activity at MIT at the time. John Ehrenfeld had just started an initiative in business and the environment there, and there was a critical mass of students. It was a very exciting time, right at the beginning when this was all brand new.

HB: How many other people were in the program?

AH: Well, there was no specific program. John taught in something called the Program on Technology, Business and the Environment, but as far as my degree went, I just made it up. MIT is a very entrepreneurial environment, and they'll let you do that.

So I made up a dual degree. There were a number of other PhD students who were interested in the topic at the same time at various schools, and the Technology, Business and Environment program was developing a Master's program, so a lot of students were interested in that too. John had a lot of energy around him. It was really exciting.

HB: When I think about civil engineering, I'm thinking bridges and all that—you know, the usual sort of stuff—but this seems completely removed from that. In what sense was this a standard civil engineering program?

AH: Well, usually a school will have an engineering management program within their engineering school. Often it's in Industrial Engineering, but at MIT it was in Civil Engineering. So, I was actually in the Construction Management branch of Civil Engineering.

Questions for Discussion:

1. What is "Love Canal"?

2. Might there be disadvantages to being in the type of "entrepreneurial environment" that Andy describes his MIT graduate program as being? If so, what might they be?

II. Environmental Evolution

Fringe and mainstream

HB: One of the reasons I wanted to talk to you sprang from a sense of frustration that I've long had, and I'm sure a lot of other people have also had.

I think it's changing a little bit now, but for a long time there was this image that there are two basic types of people. There were business types who lived in "the real world" and saying things like, *"We're living in a free-market world where we have to exercise our right to be entrepreneurial and create wealth."* And then you had people who were worried about the environment.

In other words, there was this very polarized distribution: you had the tree-huggers on one side and the business people on the other.

The business guys were busy raping the environment at any cost whatsoever to make a buck, while the tree-huggers were incredibly economically ignorant, sometimes even actively lobbying for things that might distort the economy in such a significant way that it might have negative repercussions for the environment.

This was how the debate was framed when I was younger, and it was very frustrating to watch, as there seemed to be so little common ground. I kept thinking to myself, *Can't we somehow get past this to bridge the gap between these two polarities that is obviously in everyone's interest?*

And when I look at much of your work, my interpretation, as a non-specialist, is that here, finally, is someone who is genuinely trying to constructively find some consensus, some sense of understanding, across this divide.

Is that the way you look at yourself? Or is that somehow too simplistic a picture?

AH: No, I think that's accurate. But the extreme voices are still there, and they have a significant impact. I think of the environmental movement now as being split between what might be called "bright greens" and "dark greens".

The "bright green" environmental groups see the market as a solution and business as an ally, while the "dark greens" look at business as the enemy and the market as the problem. And they're both necessary to create energy, to get change to happen.

There's an idea from political science that Herbert Haines pioneered called the "radical flank effect", which he developed around the Civil Rights Movement.

Here's the idea: Martin Luther King comes along. And his message to White America is considered far too extreme until Malcolm X comes along and pulls the flank further out—that's the radical flank effect. So now, after Malcolm X, Martin Luther King is seen as a moderate.

Similarly, in the environmental movement, those protesters protesting pipelines, logging, and so forth, are still critically necessary to stake the flag in the terrain. Then the more consensus-oriented organizations can try to work with the market system to pull it in a direction that will promote positive change.

HB: But if I was a militant environmentalist and I took this seriously, wouldn't I want to get a whole bunch of people together who were even more radical, crazy and militant than I was to push out that flank even more? Wouldn't my attitude be something like, "*You think I'm extreme? Give me a year, and you'll think I'm a moderate compared to these other guys!*"

AH: Well, there's a limit to how far you can push it before you start to get into what's considered illegitimate. Earth Liberation Front and Earth First! are terrorist groups according to the FBI. So you can push it too far, you can start spiking trees in forests and start hurting and even killing people. That can create a negative flank effect where people will say, "*You guys who burn down chalets in Aspen and free minks in mink farms, you're all crazy, so I'm not going to listen to any of you.*"

HB: How has your own work evolved throughout the years? I'd like you to trace out some sort of path, not only in your thinking, but also the scholarly reception to your thinking—and how that, in turn, has influenced the path you've taken.

AH: Well, I would answer that in two ways.

When I first started doing this, it wasn't terribly accepted in business schools. I was actually turned down in 1995 for a job at a top-tier business school where they said, "*We really love your stuff on organizational theory, but we think you're too focused on the environment.*" People were nervous. Are you an advocate or are you a serious academic?

Now the environment is a legitimate domain of research. Which is a positive step.

Within the business world when I first started doing this, I was teaching students to go into companies and teach them that the environment has strategic aspects to it, both negative and positive. Now students are going in and finding career paths available to them, and they're even starting to say, "*I don't want to go into companies and teach them how to do this, I want to do this myself. I want to start my own company.*" It's changed so much.

A second dimension I would add to this is that environment, and sustainability more broadly, has really gone mainstream: you have chief sustainability officers, environmental annual reports, socially responsible investing. It's all there, but the problems continue to get worse.

So it's now time to discover "Sustainability 2.0": where do we have to go next? There's been change to a certain point, but the problems continue to get worse, and even more radical shifts are called for.

But capitalism is a very malleable system, and it changes through fits and starts. We've gone through what I call three waves of integrating environmental concerns into capitalism: first, around 1970 with regulation, then around 1990 with being strategic, now, around sustainability. That will drop off but then things will pick up again as we go through the next iteration or shift.

Questions for Discussion:

1. Why do you think environmental issues have become much more "mainstream" in the past 20 years? Do you think that they will become even more appreciated in the next 20 years?

2. Do you agree with Andy that "capitalism is a very malleable system"? How might it be argued otherwise?

III. Beyond Punditry

The cultural backdrop to climate change

HB: I'd like to look a bit more closely at your research and scholarship now. I read a recent article of yours where you discuss climate change in terms of its cultural component, rather than specifically looking at the science per se. What were you trying to accomplish there?

AH: Well, the article is pointing out that the public debate over climate change right now is not about CO_2 or climate models: it's about world-views and values; it's about beliefs.

The central question on climate change is a cultural issue: *Do you believe that we as a species have grown to such numbers, and our technology to such power, that we can alter the global climate and might even have a responsibility to manage it?* That's a tremendous shift in an individual's conception of himself, of the world around him and his place within it. It's massive.

In my work, I'm examining the cultural dimensions of the debate. What do people hear when they hear "climate change"? What buttons does it press? For some people the answer to that question challenges the notion of God and divine providence: *"We're not in charge out there, that's hubris, we're too insignificant, we're too small: God's in charge out there, and how dare you say otherwise!"*

That's one example of how the cultural dimension of this issue emerges. One of the strongest correlates with someone's belief in climate change in the United States is his political party affiliation. That tells me that this is a cultural issue. It's not that there is better science education on one side of the political divide than the other. It's the world-views that are developed within that political ideology that makes them accept or reject the science.

HB: I want to talk about the United States a little bit later on because I think it's an outlier on this issue, as it is on so many other cultural aspects.

AH: An outlier only in the sense that there are climate sceptics in Europe but it just doesn't map neatly onto the political landscape. Therefore they're not as easily identifiable as a block.

HB: Well, there's the question of the political landscape, but I think that it's deeper than that. I'm happy to have a debate with you about this later (see Chapter 5 below). But before I get there let me just try to clarify things a bit—as, in fact, you do in this article I just mentioned.

You talk about the different questions that have to be asked when people start throwing around terms like "global warming"—there are all sorts of different associations that people have with the concepts involved. There's the matter of establishing if global mean temperatures are actually rising. There's the question of whether or not that can, and should, be attributable to man-made activities. Then there are the follow-up questions of what can be done about them, or what programs can be invoked—the need for regulation and so forth.

So there's a very clear delineation, as you say in your article, about which questions are strictly scientific and can't really be debated and which ones are less clear. You can't debate the fact that temperatures are going up, you can't debate the fact that there have been systematic studies to demonstrate this. But you *can* certainly debate whether or not it's sufficiently unique, what caused it, what we should do about it, and so forth.

So we begin with recognizing what is unequivocally true and then examine where the cultural aspects fit on top of that. Is that a fair assessment of your approach?

AH: Yes.

HB: My sense is that this paper is a call to arms, effectively saying, *"There's a bunch of stuff we know scientifically, and what we really need*

now in order to move forwards are some people in the social sciences
to step up and take a public leadership role."

And reading this, I think to myself: *OK, how do we pick those*
particular people—who are these people, exactly, who are going to
actually step up and do that? And how are they actually going to do it?

AH: There's a number of dimensions to your question, but let me
go to the one you mentioned at the end first, because it's a very
interesting one to me.

To have social scientists looking at this is important. However,
that's a different question than whether or not they should get
involved in the debate. I do think we need more studies to understand the sociology and the psychology by which people accept or
reject science, whether we're talking about climate change, nanotech, nuclear power, or any other of a whole host of issues. This isn't
something that's just restricted to one side of the political divide.

Right now, it's the political right that challenges the science on
climate change, but if you look at the GMO (genetically modified
organisms) debate, for example, it's the left that is sceptical, believing that science is currently underestimating the potential harm
involved there.

The question of academics getting involved in public debate is
a very interesting one for me. Right now we're not trained to do it,
we're not given incentives to do it, and it's dangerous terrain. If you
go out and engage in a public debate, you can find yourself crossing
a line: we're academics, we're supposed to be sources of objective
knowledge, not advocates.

So we should be out there "presenting information", but once
you get pulled into the public debate it's very hard to control your
message and control how you present it. And if you do get involved
in the public debate, it's a very interesting question of what to do if
the science shifts. How do you pivot after you've been out front in
the public? It's a very interesting question.

I think, from a social science point of view, we should be thinking carefully about how academics in all domains get involved with

public debates and step outside the ivory tower. There are a lot of challenges from outside asking universities to explain more of what their benefit is to society, and this is one way to do it. One reason why the societal debate is so dumbed-down is that not enough academics are stepping in: they are ceding the space to more biased positions.

But we need to train academics how to do it; they need to learn how to do it. It's not intuitive. Some people are not very good at it, and it's something you have to be very careful of.

Paul Krugman, Jeffrey Sachs, are they academics?

HB: They used to be.

AH: Yes, they used to be, but I don't think they are any more. Now they're pundits.

Questions for Discussion:

1. Do you agree with Andy's claim that, "It's not that there is better science education on one side of the political divide than the other"? Is that still true today? Was it true ten years ago?

2. Is the societal debate over climate change, and other scientifically-related issues, "dumbed down"? If so, whose fault is that, exactly?

3. Is there a problem with "pundits" today? What are the requirements, exactly, for being a "pundit"?

IV. Fostering Debate

Engaging, responsibly

HB: From my perspective, one of the essential things that people like you should be doing—as you are, in fact, doing—is not so much taking specific positions, but strongly advocating that there should be a debate.

That is, in my view there's this meta-issue that needs to be pointed out: *Hey, look: we're not actually having a debate. We're not actually having a rational, objective discourse about what the facts are. All we're doing is talking past each other and clinging to these ridiculously simplistic positions.*

That's not the way to solve anything, regardless of what the truth is, and regardless of what we believe. That's a message, I think, that has to get out.

You're trying to contribute to that, but do you see other people contributing to that as well?

AH: Yes. There are some academics out there who are trying to get more involved in the public debate. There's a group out of Stanford called the Leopold Leadership Fellows that is focused around the idea of getting academics out in the public forum: finding young, aggressive academics who want to have more of a say in the issues of public concern.

HB: Oh, God! Getting young, aggressive academics to have more of a say? That's all we need!

AH: Well, maybe "aggressive" was the wrong word. What I mean to say is that these are people who want to use their knowledge and experience to make a real difference in the world, not just measure their life's impact through their academic publications and citation records.

HB: OK. But my sense is that a core question when it comes to an issue like climate change, not just among academics but amongst the general public, is, *Who are you going to trust?*

The world seems divided into those loudly attacking the climate-change-deniers as greedy, dangerous, unscientific idiots who are sacrificing our global future for their short-term gain, and those on the other side who are angrily denouncing the first group as irresponsible scaremongers perversely intent on deliberately crippling our economy with their outlandish apocalyptic scenarios.

Meanwhile, for the general guy out there who's not a scientist, who's not a climate specialist, there's this sense of, *"OK, I hear all this stuff, and all I know is whether or not it has been a particularly stormy spring. I don't really know who to believe."*

So, when I hear you say, *"These academics should get out there, they don't do a good enough job interacting with the real world, they're stuck in their ivory towers,"* I'm wondering, do we really need that? I mean isn't that the *last* thing that this poor, non-specialist guy out there needs? Now there are thousands of guys from Stanford banging on his door, as well.

AH: Well, it comes down to a question of whether the scientific community has any credibility with the general public. A scientist from Stanford, the National Academies of Sciences, or the American Association of the Advancement of Sciences, they *do* carry some legitimacy.

We're dealing with a lot of static right now. A lot of people are trying to get their head around this idea of climate change. It *will* settle out. The weight of scientific evidence and the weight of the scientific agencies is massive. They are pushing in this direction saying, *"Folks, this is really happening."*

So I do think that will prevail, and I do think that we're in the midst of a generational shift. "Aggressive" was probably the wrong word, but there are young academics who want to contribute to the real world and not just the scholarly community. There are young academics out there saying, "*I don't want to just contribute to theory; I actually want to see some use from my work, some practical use of my work.*" That's happening more and more. And we have to figure out how they can best do that.

HB: What do you think? How can we harness them coherently?

AH: Well, there are so many forces going on at the same time. Social media is opening up the game—the idea that academic journals represent some private conversations that simply arrive at some fixed conclusion is increasingly outdated. Information is really becoming a public good.

Because of the blogosphere, because of the web, you can self-publish, you can write a paper that comes out in *Nature*; and then the Cato Institute or the American Enterprise Institute or anyone else can come out with a response, while the public can weigh each against the other.

Right now, in academia, we're trying to weigh this new terrain, this new landscape, to figure out how we fit into it. *What is our role? How do we show people the rigour of our work?* That is critically important. We have peer review, we have rigorous standards. All of that does give us some legitimacy.

Questions for Discussion:

1. Should scientific bodies, such as the National Academies of Sciences, be involved in making pronouncements on public-policy issues such a climate change?

2. Do you feel pressured into supporting a particular view on a pressing societal issue for which you feel you don't have enough information or expertise?

V. American Exceptionalism?

Discussions on uniqueness

HB: I'd like to get back to the comment I made before about the United States being a bit of an outlier.

In the real political world in the United States, there is this very clear fault line between the Democrats and the Republicans, as you and many others have pointed out: the overwhelming proportion of climate-change-deniers are associated with the Republican Party, while the overwhelming number of climate-change advocates are in the Democratic party.

And my sense is that not only is this one-to–one correspondence between climate change and political affiliation different in other countries, but there is also a general and much broader acceptance of the scientific evidence of climate change in those other countries compared to the United States. Why do you think that is?

AH: Well, this is where I get to the cultural components of these issues. In the United States some people, when they hear "climate change", hear 'more government": you're going to have a carbon tax or a carbon price of some sort, some sort of intrusive government program—

HB: OK, but hang on: talking about ways to deal with a problem, is different from the acknowledgement that it actually exists in the first place.

AH: But they're not separate. If I talk to you about a problem, and you say, *"That leads towards some solution that I don't want any part* *of,*" it's very hard to separate the two. You can't accept a problem and then reject the very obvious solution.

HB: OK. But people in Europe or Canada or Australia don't want more taxes either. I mean, nobody I know actually *wants* to pay more taxes.

AH: There's a visceral debate right now in this country about the role of government; there's this pressure to downsize the government, with many people claiming that it's gotten too big, too involved in our lives. That's a raging debate, and the climate change debate has gotten caught up in that.

HB: Well, I see that; and I see that it's clearly a very important factor. But I guess what puzzles me as an outside observer is this sense of unwillingness to accept the scientific realities that some shockingly large proportion of Americans seem to have that other people in other places don't.

And that's been pointed out by Republicans like Jon Huntsman who said, not too long ago, something like, "If we keep this up, we're going to be looked at as the anti-science party..."

So all this bemuses me: it's another aspect of American exceptionalism. For me, America is this very, very weird place where you have this incredibly large amount of people who are completely ignorant of the most basic scientific processes, it seems substantially higher than anywhere else in the developed world.

AH: The California Academy of Science did a study and found that something like 75% of Americans don't understand the scientific process.

HB: Right. It's not that everyone is so wonderfully scientifically literate and cultivated and "children of the Enlightenment" everywhere else, but relatively speaking, America's very different—clearly, objectively worse.

So there's that. But then on the other hand, the US is this focal point of remarkable scholarship, remarkable innovation, remarkable technology. An overwhelming proportion of the world's best research institutions are found in this country by any objective measure.

If you take the United States out of the equation in terms of scientific research and innovation, it's nothing short of a global disaster. And not just for science. Simply put, it is overwhelmingly the most important country for research and scholarship, across pretty much all disciplines.

So that's wonderful, and hugely important. But then you're left wondering how on earth it's possible that in the teeth of so much scholarship, so much innovation, so much intellectual excellence that is really, to a large extent, driving the entire planet forwards, the average guy on the street in this country is such a complete and total ignoramus? I mean, it's just so incredibly odd.

AH: I don't know how to answer that except to say that we're a very heterogeneous place. One element of the climate-change debate is distrust of science, and that's a strain that's been in the American society for a long time.

But at the end of the day the majority of Americans *do* trust science. They do. If they get sick, they're going to a hospital. And what's that hospital based on? The scientific process. They get in their car, they help their kids with their chemistry or biology homework.

The idea that on this one issue science got it wrong, really will not stick. Because the majority of Americans *do* believe in science, they *do* trust the scientific method.

HB: But you just told me there was this study that said that 75% of Americans don't understand science—

AH: There's a difference between understanding, between passing a test, and trusting it. If they get sick and the doctor says "*We're going to start chemo,*" they're not going to look critically at the scientific papers. They're going to say, "*I trust that doctor. I'm going to trust the process that got to that conclusion, and we should take this step.*"

HB: And similarly, on a much more mundane level, they'll turn on a light switch and expect the light to actually work.

AH: Yes, and turn on the car.

HB: OK. But there are still many political-cultural aspects of the situation that are very troubling. When you watch Republican Party conventions—I mean, this is not exactly a fringe party—you see people in major political positions of power in the most dominant country on the planet who are saying all sorts of things that you are cringing at. It's pretty alarming for all sorts of reasons, not least because if we're going to do anything significant to address climate change, it's obviously got to involve global regulations. It's a global problem.

AH: You just hit on another hot-button issue. It's clear that the Republican Party has been pulled further to the right: libertarian, evangelical, protection of national sovereignty (that's what the immigration debate is all about) and so forth. Once you start talking about global governance, you've just hit a hot-button issue for the far right. They don't want that.

HB: Well, some of them don't even seem to want a national government. They don't even want local government.

AH: They have very strong views on all of that. And the point is that these are the hot-button issues that those people think about when they hear the words "climate change".

Jonathan Haidt does a lot of nice work on the psychology behind decision-making. When you're faced with a difficult situation, your emotions kick in first, and reason kicks in second.

Some people hear the words "climate change" and they just instantly think: "more government, loss of national sovereignty, the UN is not trusted." Then they look for reasoning to support that position.

Now, some may get upset that I said that. They might say that they actually look at the science, and their decision about climate change is based on the actual science. But the science is quite compelling that this is a real effect. So what are they seeing that causes them to reject the science that is out there? Distrust of environmentalists, distrust of Democratic politicians, distrust of scientists, fear of big government.

They might fall back on the notion that there is no religious mandate to protect the environment, to protect the global climate: *"God will take care of things—he promised Noah that he would never flood the Earth again. The Genesis mandate says that we are stewards of the environment: it is there for our use, and it doesn't have inherent value."*

Meanwhile, catastrophic scenarios are rejected out of hand. Take the movie *The Day After Tomorrow*, the idea of Manhattan being underwater, with glaciers going down Madison Avenue. People see that and say, *"Nonsense. Once again, it's the environmental movement saying that the sky is falling."*

All of these tumblers start to fall into place, making some people retort, *"Here they go again. They're anti-development, they're trying to roll us back, they don't want us to develop."*

I actually think that it's unfortunate that climate change has been coded entirely as an environmental issue. It's really not. It's a scientific issue, it's a social issue, it's an economic issue. That will come out over time, but right now it's coded almost entirely as an environmental issue.

Who's leading the charge? It's environmental groups, it's people like Bill McKibben.

There have been others who have tried to push it in a different direction, saying, *"No, this is an issue of economic competitiveness,"* or, *"This is an issue of national security,"* or *"This is an issue of technological development."* But for now it's still regarded primarily as an environmental matter.

Questions for Discussion:

1. *To what extent is the notion of "coding something as an environmental issue" part of the problem? Who is doing the coding and how are the rest of us allowing such "coding" to occur?*

2. *Is there a problem with the decision procedure described by Jonathan Haidt that Andy alludes to in this chapter? In what ways is it reasonable to accept that people should make decisions this way? Is this the way that people make decisions everywhere?*

3. *Do you think that the apocalyptic scenarios that some environmental groups have invoked have backfired? If so, what do you believe they should say now to rectify matters?*

4. *Do you share Howard's views that the United States is strikingly different than most of the developed world, or do you think he is misinterpreting matters?*

VI. Talking the Talk

Communicating science better

HB: You make an analogy to the issue of smoking and the tobacco industry in terms of the mechanics of changing public perceptions—that in the case of smoking, for a long time there was very significant scientific evidence, but it took a while until it caught on and people started recognizing it and moving beyond the influence of powerful interest groups that were in place.

So is it just a question of waiting again?

AH: No, no. Social change isn't a linear process, and it can go through sudden shifts. It needs advocates, it needs social entrepreneurs to try to push it, taking advantage of events as they emerge. Hurricane Sandy was just such a critical event.

Let's compare the two hurricanes Sandy and Katrina. One created change, one didn't. Both were massive storms that whacked a major urban centre. One hit a minority, politically disconnected, poor population, while the other hit a white, politically connected, affluent population. One didn't have a national spokesperson to bring it on the international stage. The other had Michael Bloomberg. Right before the presidential election the cover of Bloomberg Businessweek said, *It's Global Warming, Stupid.* It was framed as climate change. People had also tried to connect Katrina to it, but that didn't work. This time it did.

That is reflective of the idea of social entrepreneurs stepping forward, taking advantage of an event. People have mocked Rahm Emanuel, when he said, *"Never waste a good crisis."* But that is sound social theory: people change in the face of something that disrupts their common beliefs.

HB: Well, it's sound tactics. He probably shouldn't have said it quite so brazenly—

AH: Right.

HB: —but that's a whole different issue. But here's what makes me squeamish when I hear you talking about social entrepreneurs using the crisis, leveraging it.

As somebody with a scientific disposition, I'm very comfortable with highlighting the scientific consensus that's been established on climate change, or anything else for that matter. I'm very comfortable about the fact that a clear pattern has emerged based upon all sorts of data about what has happened over a long period of time. By and large there seems to be a clear indication of what's on the more speculative side, what's on the more established side, and so forth.

But once people start looking at individual data points, like, *"Oh, there was a really big storm—something must be going on!"* then I get nervous. Because then you face the danger of someone else coming along and saying, *"Well, there's no climate change because it was really cold this winter"*, and all that kind of silliness.

I understand what people are trying to do: they're using an event to get attention and get people to focus on the bigger picture. But isn't there a risk of jeopardizing the scientific process by doing that?

AH: It depends on how it's done. Scientists stepped forward right after Sandy and said, *"This storm is not climate change. This storm was not created by climate change, but climate change created the conditions by which it was more extreme than expected—the ocean was warmer and so forth."*

Weather is not climate, you're absolutely right, but what they're trying to do is get the process going so that people are open to the issue. People respond to what is salient and personal. That's why polar bears sell and snail darters don't: it's charismatic megafauna, people have this affinity to it: it pulls the heartstrings.

There are studies that show that people who have been exposed to extreme weather events are more inclined to believe climate change is real because they can accept that the environment can turn nasty on them, that it can become hostile.

HB: OK, but I don't want to convince them that way. I want to convince them by educating them about science.

AH: Well, you start with the data of a weather event, but then you quickly transition and say, *"This is not climate change. Climate change is about long-term trends in global mean temperatures. It's about broad-scale shifts over a longer period of time."* But this is the door opening.

This is a communication effort that a lot of scientists get upset over, as you are right now, but scientists need to recognize that you might have the right idea, you might have the right answer, but now you've got to convince people that it's the right answer.

Scientists who think that, *"I just need to come up with my right answer and people are going to accept it"* while ignoring the social and political context of what they're doing are really missing the point.

Climate change really did not become a contested issue until the Kyoto Protocol started to come into form, when it started to threaten some very powerful economic and political interests.

They were able to mobilize in order to get this movement to say, *"No, it's not happening."* In business parlance I describe climate change as a market shift. And in a market shift you're going to have winners and losers; and the losers will resist the market shift.

That's what we're watching right now. And they're funding efforts to debunk the science: that's a critical part of the debate in the United States.

HB: I completely agree that you can't just say, *"Oh well, we're just going to present scientific data and if these people are scientifically illiterate then to hell with them."*

But I've been angered by many of the climate change advocates for a long period of time, because there's a lot of rhetoric from people who are associating themselves with science who are throwing stuff out that isn't necessarily true, or they're using all sorts of *ad hominem* attacks: *"We're a part of the educated elite, we're the scientific guys so we understand these things, and anybody who doesn't is just some sort of knuckle-dragging cretin who clearly doesn't know anything and should be ignored."*

Not only is that counter-productive towards moving forwards constructively, but sometimes it's doing more of a disservice to science, because along the way you might be framing things incorrectly, stating all sorts of untruths.

AH: But that's what happens when any kind of scientific issue gets thrown into the public sphere: different people take up the charge and become associated with it. Right now the environmental movement has been the leading force in climate change and some have gone beyond where they should, as the other side has gone beyond where it should.

How do we get credible scientific information into the public in a way that doesn't get distorted by the political aims of those who are purporting it or rejecting it? There's the question.

HB: So how do we? What should we do?

AH: Well, I would like to see more scientists getting involved in the debates and speak for themselves: present the evidence in a way that's carefully constructed.

That's a conversation that's taking place in the academy right now: Roger Pielke wrote a book called *The Honest Broker*. It's a really nice book on what is the role of the academic in policy debates. He talks about the "honest broker" as somebody who brings forward and lays out all the work that's out there, letting the social and political process work based on that.

Some people agree with that and some people don't. Some people think that scientists should come out and say, *"Let's put some percentages on this: here's what the science tells us, here's the answer. We're not going to confuse things with all the other distractions from the consensus statement."*

It's still being worked out how scientists can communicate, but we need to, because we live in an increasingly technological age and we're in a functioning democracy. People are voting on things like nuclear power, GMOs, nanotech, health care, gun control: issues that require good data and good analysis. We need to figure out how we bring that data and analysis into the public debate in a serious way.

Questions for Discussion:

1. Do you agree with Andy that scientists should take a more active role in the public communication of climate change? If so, how do you think that they should do so more effectively? If not, why not?

2. Is Andy implying that scientists are generally objective and apolitical? Under what circumstances can we be certain that "credible scientific information" can be separated from a surrounding political climate?

VII. Preaching to the Choir?

How to make genuine social progress

HB: But when I hear you talk about getting more scientists involved, my sceptometer starts going up.

AH: Why?

HB: Well, I can't help wondering, *Who's going to listen? Who's going to listen to passionate, articulate, well-motivated scientists?* And my guess is that the answer is simply: those who are already predisposed to doing so.

But the people *I'm* worried about connecting with are those you were mentioning a minute ago: those who are absorbing a different sort of rhetoric, the ones who might be insecure for whatever reason, the ones who are inclined to position issues like climate change or emerging technologies in terms of winners and losers in the market, naturally fearful that they might be one of the losers. In other words, I'm concerned about how to broaden the message to go well beyond the already converted.

I get this sense that having more scientists get up and talk just adds to the same old choir. I'm wondering how you can really move out and foster broader debate with people who might have very different views. I see that some of your work is speaking exactly to that, head-on. But getting another bushel of people from the National Academy of Sciences out there? I'm not convinced that such an approach, in and of itself, is going to change anything.

AH: Well, what I'm saying in my work is that the spokesmen that people will respond to are those who are a part of their referent group, their cultural community, their tribe. An evangelical is going to listen to an evangelical more than he may listen to the National Academies of Sciences, so we need more evangelicals speaking on this issue. We need business people, we need politicians, we need people from groups that people trust, and we need it at the local level.

People need to hear it at the Kiwanis Club, in the golf leagues, at the town hall. It has to be not just a top-down movement, which it has largely been so far, but a bottom-up one.

That's where I think Bill McKibben has actually been able to do something. He's created a grassroots movement. I have questions about his endgame, but he's been able to create a constituency around this issue involving young people. He's been able to turn it into an issue of social equity: *your world is going to be damaged by what we're doing now and you're going to have to live with it.*

He's been able to mobilize many people. But I'd like to see even more people making new connections. I'd like to see more anglers saying, "*You know what, this is going to ruin the habitat for the environment that I enjoy*", or others saying, "*Holy smokes, Michigan just lost 90% of their cherry crop last year because of some very strange weather. Scientists are saying it's climate change...this is bad.*" If they start to connect things to their own personal interests, then you'll get change. Then you'll get people moving on it.

HB: Apropos of that, I've long been confused by the typical demographics of environmental issues. There are an awful lot of hunters and fishermen who statistically seem to be associated with the Republican Party, but whom you'd imagine would naturally be extremely keen environmentalists.

I mean, these people live in a world where a healthy natural environment plays a much more preeminent role than your average Democratic supporter in Manhattan.

So it's always been a bit confusing to me that environmental issues get portrayed as something naturally associated with hip, young, progressive urban, leftists as opposed to sceptical, retrograde, rural types, because you'd expect that it would be those who live directly with the environment, who have an intimate relationship with it, who would be the most passionate about protecting it. It's perplexing to me.

AH: Well, we have to be careful of these broad categorizations. Not all Republicans or Democrats feel the same about all issues. And if we rely on these stereotypes too much, it starts becoming a self-fulfilling prophesy: if you're a Republican, you must not believe in climate change.

There are plenty of instances of anglers and hunters advocating very strongly for habitat protection while working alongside environmental groups. It's not an uncommon alliance.

Questions for Discussion:

1. Why do you think that a statistically large proportion of evangelical Christians are sceptical of climate change?

2. To what extent do you think that opinions on climate change in the United States are correlated with political affiliation?

VIII. Energy Renaissance

Government's role

HB: Before we move on to *Flourishing*, I'd like to briefly get back to this idea of harnessing the current crisis to drive innovation. I've heard President Obama talk on several occasions about the need to look at new and emerging technologies together with environmental issues as an opportunity upon which American innovation can flower.

This strikes me as quite a reasonable thing to be thinking, because, as I said before, unquestionably one of the most impressive things about America is its ability to innovate, its ability to marry scientific thinking and entrepreneurship in a very substantive way.

Is this message getting through at all? It seems reasonable to me that there should be a kind of repositioning: *we're going to take this as an opportunity to innovate and create something new*. There's a lot of money to be made if you do environmentally creative, interesting things.

Is that spirit generally being adopted right now, or not so much?

AH: At the risk of hyperbole, we're in the midst of an energy renaissance. We're shifting right now. It's going to be so different 30 years from now.

Let's start with the power grid. The grid's a joke in this country: it's falling apart. We're going to spend over a trillion dollars improving the grid in the next 30 years. Is it going to be the same grid we have now? Absolutely not. Will it start taking advantage of smart grid technology? Will we start to unify the grid to be able to get energy from where it's created to the demand loads in urban centers? What will be the future of distributed energy? And what will actually make parts of the grid less important? Energy independence from the grid is actually important to many people.

There are appliance manufacturers who have appliances ready to go that can actually talk to the grid and turn on when energy is cheapest, if we have real-time pricing. Demand management is very strong, and different kinds of energy sources are out there.

You can go to an auto dealer right now and buy one of many drivetrains, whether it's hyper-efficient diesel, improved internal combustion engine, hybrids, or electrics. When will fuel cells come online? There's a lot of research on that. There's a lot of research on battery storage. Once that's cracked you're going to watch the automotive sector shift.

Meanwhile, there's a lot of discussion about the car of the future. We're moving towards cars that can actually drive themselves. It sounds like science fiction, but think of what information technology can do to the smart home: people's awareness of their energy bill is increasing and that will drive behaviour change.

All these things are happening around us, and a lot of the political debate gets hung up on isolated cases of mismanagement and failed government loans like what happened with Solyndra—that's all they see. There's a lot more innovation going on, a lot more exciting stuff going on. And again, where we'll be in 30 years is going to be so different.

HB: So how can this be made to happen faster? There are real reasons for optimism, which is obviously what you're saying, but at the same time there are all these cultural issues that are dragging us down. How can we move in a faster direction?

AH: Well, it would help if we had an electorate that was more supportive of these issues so that their politicians would support shifts in the tax code that could move beyond the stale and bizarre idea that any kind of tax structure or subsidy is some kind of an intrusion on the market.

The market is a man-made set of institutions. The government sets the rules. We can't price fix, we can't collude, but suddenly by having subsidies for solar companies, that's somehow the government picking winners and losers, which is unacceptable. Meanwhile, people say, *"Don't touch my tax credit for my home mortgage!"* which amounts to exactly the same thing.

It's important to appreciate that the government already does this sort of thing. And if we see the future in a particular area, it's a smart government that will push in that direction.

But now, we're getting into the cultural debate:

"OK, so now you're talking about industrial policy..."

"No, we're a free economy. Don't get into industrial policy..."

That's where these hot-button issues start to emerge: people start to resist, saying, *"No, the government shouldn't be doing that."* We somehow have to get beyond that.

Questions for Discussion:

1. How do you think people thirty years from now will judge our current transition to a green economy? What will be the perceived successes and failures?

2. Are you excited or indifferent about the notion of "driverless cars"? How do you think the car of the future will differ from those today?

IX. Reinventing Sustainability

Imagining the long term

HB: So, let me move to *Flourishing* now, because you talk about the need to make significant cultural changes, or at least one of you talks about that, while you express support. I should say that this is a dialogue between you and John Ehrenfeld.

AH: That's right.

HB: He says some things that are very bold and provocative in terms of the need to change our cultural values. It's virtually utopian— much more ambitious than what we've been discussing today.

AH: Well, John was my teacher when I was a graduate student and he's been a mentor to me ever since. I see him as a visionary. I see him as looking much further out than most of us can see. He's a very deep thinker, he's been thinking about these issues for a long time and he's very philosophical.

What I think he's pointing out is where we need to go for the long term, that some things we're focusing on now are not going to take us where we need to go. Yes, you can buy a compact fluorescent light bulb and screw it in. That's great, you're reducing your energy load. But there are still a lot of materials that went into that: you're just making the production of light *less bad*. How do we shift from there to actually making our technological society *better*?

He makes the really powerful point that all our efforts right now are *reducing unsustainability*, which is a fundamentally different thing than *creating sustainability*. In our discussion, he explained it this way, which I really like. He said, "*In Iraq we've stopped the war. That's fundamentally different than creating the peace.*"

In other words, he's not against windmills and compact fluorescents and hybrid cars, because he says that's slowing the velocity at which we're heading towards that brick wall. But if we want to stop and reverse course, we have to think fundamentally differently about this problem.

Now, I don't expect people to adopt what we're talking about in this book tomorrow. But I do think he offers us a guidepost towards where we need to go in the long term. If we're going to get a grip on sustainability, we have to rethink consumption. We have to rethink a lot of the dominant values in our society, like the notion that we are defined by what we own: the bigger the house and the fancier the car, the more status you have, the more worth you have, both self-defined and defined by others. If we don't get a grip on that we're never going to get there.

To John, it pushes on the idea that fundamentally, if we are going to be sustainable, we have to rethink our dominant conceptions of what it means to be human, what our relationship is to the world around us, and how we relate to each other. It's a tall order. It's huge.

What I think he's doing is putting sustainability on par with a massive shift in our thinking akin to the Enlightenment. In the Enlightenment we moved from superstition and confusion about the world around us to a scientific, mechanistic view of how nature works.

That has brought us some great advances, but it has also brought us some real problems. How do we start to shift again in our conceptions of ourselves as a species, and our relation to the world around us? That's how big sustainability is.

To me, he's calling that out. This isn't simply just, *Adopt a couple of technologies and continue to live the way you've lived before.* We have to shift the way we think about how we live and how we relate to the world around us. It's big thinking. It's very big thinking.

HB: And it's centred on this whole issue of sustainability. If you examine the word "sustainability" closely, you'll quickly come to the notion of sustaining something, keeping it going. But if the ship is going in the wrong direction, you don't want to keep it going. You want to turn it around.

AH: The point we make in the book is that the word "sustainability" itself is about stasis: staying the course, keeping it steady. Instead, John redefines sustainability as the possibility—we don't know if we're going to do it—of humans and other life forms flourishing on Earth forever. All those pieces of the definition are critical but the word "flourishing" is about thriving: it's about growing, it's about positive dynamic change.

That's a much more attractive goal. So, should we be talking about genuine happiness and satisfaction, or are we talking about *more stuff*? I think he's trying to highlight that distinction.

Questions for Discussion:

1. How possible do you think it is for our contemporary society to transition to one that is less consumption oriented?

2. Are there any places on earth right now that you would characterize as being oriented towards"creating sustainability" rather than "reducing unsustainability"?

X. Surprising Revolutionaries

Idealistic business students

HB: I'd like to explore that a little bit. But first I'd like to get a better sense of what *you* believe, because, throughout *Flourishing*, it's very clear that you're highlighting John's beliefs and views. *He's* the visionary, and *you're* having a conversation with him to flesh his ideas out in detail.

But *I'm* having a conversation with *you*, so I'd like to know how much of all of that, broadly speaking, you are sympathetic to.

AH: Well, I am sympathetic to what's in the book. At the root of it lies a change in our culture, a change in our beliefs. But I'm teaching students, and I'm talking with companies in the here and now, so my work is naturally more pragmatically focused: not what we're going to do in 50 years, but what we're going to do in the next 1, 2, 5, 10 years: *How can we best train our students to go out into businesses and develop the solutions to the problems we face?*

When I teach my students, I'm pointing out to them that business is the most powerful institution on earth. If business isn't developing solutions for our social and environmental problems, they will not be developed. Capitalism is malleable and it is shifting. There are multiple signals of this shift; and companies are, to less or more degrees, attentive to those signals.

I don't expect students to walk into Ford right now and say, "*You know what? Stop making cars, and start thinking about mobility as a totally different thing!*"

That's not going to happen tomorrow. I'm also not suggesting walking into ExxonMobil and saying, "*Stop producing oil!*"

These kinds of changes will take time. But starting to be part of a broader process, starting to shift the kinds of products, processes and solutions that companies provide, that's what I'm focused on as a teacher.

HB: I have to admit that this is all a bit confusing to me. I'm going to say something that might sound provocative, but I don't particularly mean it that way. I enjoyed the book very much—I thought it was inspiring, and I certainly very much concur with this notion of having to get beyond defining yourself through the stuff you have: this sort of silly, rampaging consumerism that I think to some extent afflicts the culture of all Western market economies.

But at the same time there's a recognition that we're not going to sit around and pretend that we're living in some Buddhist monastery somewhere. Market economics and capitalism is not only important for us, it's also a tremendous driver to move all sorts of people out of poverty. Take, China. Obviously, we're not going to go to them and suddenly say, "*Guess what? We've changed our minds. You shouldn't be actually trying to have things after all.*"

I understand the idea that we have to be working, pragmatically, within a capitalist framework, to try to somehow establish a balance.

So while John is calling for a sea change in our cultural attitudes, you're talking about how we need to get beyond stereotypes and move forwards to develop meaningful cultural dialogues. And the driving force behind all of this is the notion that the whole idea of *sustainability* up until now has been just sort of fiddling around the edges with stuff, and we have to progress to the next level of awareness and meaning in order to generate genuine positive change.

As I said, I'm basically fine with all of this. And yet—and this is my confusion and potentially my provocation—from my perspective, when I imagine a movement to try to change the hearts and minds of people towards a less consumer-oriented society, pretty well the *last* place I would think it would come from would be a business school.

My impression is that most of those people are focused on money, well-paying jobs, material possessions, and so forth. Which is fine, of course, as far as it goes—but to me they're hardly role models for generating broader cultural shifts away from materialism. Instead I would look at people studying things like philosophy, Renaissance history, cosmology, that sort of thing—*those* are the people I would imagine would have more resonance with these sorts of ideas than the average business school student.

AH: Well, there's definitely a demographic there like you describe, but more and more students are coming into business schools because they want to make a positive change in the world, and they see that business has the power base to do it. They see the opportunities.

I told you that when I first got into this I wanted to try to teach students to go into companies and help them to see environmental issues as strategic opportunities. Now we have more and more students coming out saying, "*I don't want to go into a company and teach them; I want to do it myself.*"

Many of our students are creating start-ups. Increasingly, young people are motivated by the idea of creating a company that can try to address social and environmental issues.

There's also a focus now on the "hybrid organization", the sort of organization that lives in the blurry space between the for-profit and nonprofit world—while you're here in Ann Arbor, go check out Ten Thousand Villages, a nonprofit using a for-profit model.

Meanwhile, you have for-profits that develop very strong social and environmental missions. We're seeing more and more students who want to be involved in these sorts of initiatives. They want to make a positive impact on the world, and they see a business as a way to do it.

HB: So business schools are really attracting people these days who are idealistically motivated?

AH: Some. Not across the board.

HB: Have I been missing the boat here? Has this been going on for a long time? Or has it changed relatively recently?

AH: It's changing. There are student organizations that are geared around that impact, the business of social responsibility. Here at the University of Michigan we have the Erb Institute, a joint program between the Ross School of Business and the School of Environment. In three years, students get an MBA and a Master's from the School of Natural Resources and Environment. It's been here since 1995, and the number of students is in a steady upward trajectory. We graduate about 30 a year now and we have about 400 alumni. These are students who really want to make change. They want to change the world, and they see a business as a way to do it.

They don't see the objectives of business and the objectives of protecting the environment as being in opposition. We're not going to develop the solutions as long as they're set up in opposition.

When I teach these issues in a business school, I don't teach corporate social responsibility. If I were to do that, what I'd be telling students would be, *"Go into your other classes, learn how to maximize return on investment, and then come into my class and I'll teach you a different value set."*

That's not sustainable. The first step towards having a sustainable business is that it makes money. The key question is: *How do you do that in a way that can also accomplish some environmental and social goals?*

Of course, there are some limits: policy is necessary, but market drivers take the forms of insurance companies, investors, consumers. Up and down the supply chain there are all kinds of pressures out there that are driving companies in this direction to start attending to these issues.

Take GE's *Ecomagination*. What's happening there? Well, they're not wrapping their arms around a tree. There's an opportunity to make money by promoting technologies that are part of a carbon-constrained world.

That's exciting, that's the sort of thing that our students get jazzed by.

Questions for Discussion:

1. Can business leaders enable genuinely sustainable structural change to our contemporary beliefs and values, or are they necessarily "part of the problem"?

2. To what extent do you believe that the structures of contemporary capitalism are compatible with a profound redefinition of our notion of sustainability, as described here?

3. What role might "enlightened philanthropy" play in successfully transitioning to more environmentally-sensitive cultural values?

XI. Setting Ideals

Towards a North Star

HB: So, I get that. As I mentioned earlier, I'm strongly convinced by your argument that capitalism and environmentalism don't at all have to be at odds with one another, and I can well understand how, if I were a young person today interested in starting a company, I'd be passionate about both making money and environmental protection. That makes complete sense to me.

But what I *don't* get is how you can coherently combine that view with what John seems to be saying in *Flourishing*, how we should be redefining our cultural values so that we're not focused on constantly consuming and defining ourselves by what we own, because to me this seems inextricably tied to the standard notion of economic growth within a market economy.

Maybe I'm naive, but that's the sort of thing I wouldn't expect from somebody who's going into business school to have any sort of resonance with, nor would I expect people who are funding a business school to have any sympathy with it. So, am I wrong there?

AH: I think what John is providing is an aspiration, and I'm not sure we'll ever achieve it. We will never actually achieve a sustainable society: we'll never actually get to the point where we'll say, "*Done. We're sustainable.*" It's a continual striving towards an objective that I frankly think is not possible as an end state. But he is giving us a direction to reach towards, a North Star, if you will. We're never going to touch the North Star but it gives us a direction to move towards.

Our students come out of school today and they have to live in the here and now. I love what John is trying to say, I endorse what he's trying to say, but I don't think he's saying, *"We're going to get there tomorrow."* He's saying, *"This is where we need to reach for."*

But how we get there is going to require steps along the way. That's where my students start to come in.

HB: So, what do you hope this book will do? What are you hoping the reaction will be?

AH: Well, the hope is that it inspires, that it really takes sustainability to a different place. I do think that the word itself has become stale: it means everything to everybody, therefore nothing at all—Dan Esty at Yale said, *"Let's throw the word out."*

I'd like to give "sustainability" some cultural roots to really express how we think about ourselves and the world around us. That's what's at the root of it.

You have to change your beliefs and values. This is consistent with the work I'm doing on climate change. If you're really going to deal with climate change, we have to confront some of the world-views we possess. That, I agree with wholeheartedly.

Bringing it into that domain and giving us that long-term aspiration of what we really should be thinking about is quite exciting. I hope it will inspire people and get them to stop and think. They may not change tomorrow, but I hope it will do something to adjust the conversation and begin to shift it towards where it really needs to be.

HB: What's been the reaction so far?

AH: Well, it's only just come out, so it's too soon to say.

HB: OK, but what has the informal reaction been from your friends and colleagues? Have they said something like, *"Oh-oh, Hoffman's out on a ledge with this one..."*

 AH: No. No one's said that yet—at least, not so far as I know.

Questions for Discussion:

1. Oscar Wilde famously wrote, "A map of the world that doesn't include Utopia is not even worth glancing at...". To what extent do you think the contents of this chapter are in line with that sentiment?

*2. How do you think **Flourishing** was received throughout the business community? The environmental community? Other communities?*

XII. Impact

Changing hearts and minds

HB: I'd like to explore how we might be able to foster real impact, and I'd also like to return to the issue of global cooperation. If we look at a global phenomenon like climate change, it seems that if we want anything significant to happen, we need to have the United States to play a significant role in that global mission, that global development.

But this brings us to the important issue you were speaking of earlier, that any sort of multilateral, multinational, global-governance-type of solution is very much a political hard sell in this country. So, how can we get beyond that? How do we move towards some sort of progress on that front?

AH: Regarding global governance, it's not clear. We do have global governance in various forms, but we have to be creative. Maybe the UN isn't the right body to do this. What if it was the WTO? I don't know. But that doesn't mean the conversation stops.

HB: Of course. And, moreover, the conversation has to recognize what's worked and what hasn't: it has to be a realistic conversation not a utopian conversation. So, what would you do, exactly? What would you do if you were president of the United States?

AH: I don't know. Now we're getting a little outside of my area of expertise into international diplomacy and international politics.

HB: Fair enough. But this is the speculative part of the conversation.

AH: Well, when it comes to academics getting involved in the public debate, I think a good motto is, *"Stick to your knitting."* When I see economists giving opinions on climate science and climate scientists giving opinions on cap and trade, I cringe a little bit and say, "*Stay where you're an expert before you step into the public debate.*"

Jane Fonda should have stuck with acting and not gotten involved with nuclear power. I think the same is true with this area. Recommendations on international policy are a little tricky for me.

HB: OK, but if I'm some guy who's sitting in the middle of Wisconsin, say, and I'm concerned about these things—I'm worried about my crops, I'm worried about the economy, I'm worried about all sorts of things, as everyone is—what should I do?

I hear these calls for a cultural change. I understand that we have to make progress. I understand that we're politically at some sort of stalemate: something has to give. But what should I *do* exactly? How should I go forward?

AH: Well, first of all, you don't come out and say, "*We're going to change a culture, we're going to change your values.*" No one has that power. You change behaviour and values follow—in fact, sometimes they do and sometimes they don't. You can have a set of policies and people might start to adopt the values behind them, but sometimes they won't. We had prohibition and it was a disastrous mistake: no one accepted it.

So you try to change behaviour, and then values might follow. What are the ways to do that? How do we get people to start to think differently? I think it's starting to happen around certain technologies, certain changes. People are moving more into urban centers now, walkable cities are much more attractive than car-oriented habitats. There are some consistent shifts that are happening that foster a better lifestyle and better standard of living.

One thing that the environmentalist movement has been rightly criticized for is focusing overly on the negative: *Go **this** direction or bad things are going to happen.* A much more profound message is, *Go this way because it's a better direction to go in.*

That's another element, I think, where people's backs get up on climate change. They hear criticisms, like, "*It's your fault because you live in that big house and drive that fancy car."* People naturally get offended by that.

There's lots of very exciting stuff happening in the area of green building. Anyone who doesn't hyper-insulate his building in this day and age is out of his mind: he's throwing money out the window.

There's a really nice book by Sarah Susanka called *The Not So Big House*, which essentially says, "*Don't build this big box with these huge rooms, and then worry about decorating it: shrink it down. Use that extra money to make the space inside much more attractive, much more flexible."*

It's really cool stuff, and a much more beautiful way to live. Those are the sort of concrete measures, I think, that we all need to focus on.

HB: So are you optimistic, as a general rule?

AH: As a general rule I am, yes. You've got to be. In *Flourishing*, I ask John about this—it's one of my favourite parts of the book. We go through this little riff on the difference between hope and optimism.

Optimism is looking at the odds and saying, "*You know, the odds tell me it's going to work out."* Hope is based a little bit more on faith, saying, "*You know, whatever the odds say, I still believe it's going to work out."* So you can be pessimistic and hopeful: you know the odds are against it, but you still think it's going to work. And if I wasn't hopeful, I'd give up.

I'm hopeful because of the students I see and the younger people who really want to roll their sleeves up and get this done. David Orr describes hope as a verb with its sleeves rolled up. I really like that. I look at my students, these students in this program that I'm running where they get this dual degree in business and environment. They want to find a way towards getting business solutions to our environmental and social issues. That's exciting to me; and that's hopeful.

HB: So now it all makes sense to me: it's just **your** business students: they're the statistical outliers. All the others are the usual rapacious, consuming, corporate types.

AH: These sorts of programs are popping up all over, because students want this stuff. They really do. And business wants it. The best signal I can give you right now is that our students are starting to be recruited more and more by the top management consulting firms like McKinsey and Deloitte. They see a need for it. It's part of the business environment. It's exciting stuff.

HB: Well, sorry, but you lost me there. The fact that McKinsey and Deloitte are interested doesn't actually turn my crank...

AH: Well, they can sell it. But the students also want to start their own business or go to work for Ford Motor Company on alternative forms of mobility. What's the future of mobility? Our students are changing the conversation from, *How do we make another car?* to *How do we think about mobility?*

And it's not just Michigan. There are programs around the country, dual programs or certificate programs. They're popping up a lot.

HB: If you look at the students who are coming through your door now and compare them to those of 15 years ago, do you see more passion, more enthusiasm for making a difference in the world?

AH: I think so: more passion, and more sense of the possibilities. The students who are coming in are very excited, and they're coming in greater numbers. It's also diffusing into the rest of the population. It isn't just our students. We have these Erb Institute students here at Michigan, but the overall business environment at the Ross School of Business is strongly influenced by the Erb Institute. We work on areas like positive organizational scholarship, non-profit management, base of-the-pyramid studies. A lot of that activity diffuses throughout.

Of course, there are other schools that look at this and say, "*Not our bag,*" and that's fine. But there are plenty of schools that *are* starting to focus on this: Stanford has a strong program, as do Duke, Yale, Santa Barbara, Northwestern, MIT, and Harvard. I can go down the list: they're all developing programs in this area.

HB: And what about outside the United States?

AH: That's a great question, because it not only applies for programs like this, but also business schools in general. It used to be that the American business school was the dominant player, but there's a lot more serious competition now from Europe and Asia in the business school world. The idea of business as a social force in society is not as new, particularly in Europe, as it is here.

HB: Is competition really the right way to look at it? I constantly question these things. Sure, schools are competing for individual faculty members and students. But on the other hand, in the overall scheme of things, if you're right—and I truly hope you are—that there are all these young, dynamic, socially conscious individuals who are coming through business schools these days, then at some level a competitive model is not the way to look at things. It's in everyone's interest to have more and more of these people being as successful as possible.

AH: That's right. Both points are true. If a school calls me and asks, "*Would you come out and talk about how to develop a program like this?*" I'd gladly help them. There are no proprietary secrets here.

But by the same token, business schools *are* competing for applicants. And right now the applicant pool for business schools is flat. That's partly the present state of the economy, but partly a result of the existence of many viable business schools outside the United States. The United States is not the only game in town anymore. So in that sense there *is* a competition.

Questions for Discussion:

1. Do you share Howard's scepticism about management consultants? Do you think he is exaggerating? Not sceptical enough?

2. What role do you think global institutions should play in combating climate change? Are they necessary to enable meaningful global progress? If so, how optimistic are you that America can play an active role within that global structure?

XIII. The Passion Principle

Discovering our calling

HB: We talked earlier about your youthful experience of building houses, but at the time I didn't specifically refer to the award-winning memoir you wrote about that experience, *Builder's Apprentice*, which you must be very proud of.

So I'd like to talk about carpentry for just a moment now from a somewhat different perspective: with all your time taken up with writing papers, inspiring students and trying to change society, do you miss just working with your hands sometimes, being immersed in an environment where you can be at peace with yourself and see the tangible fruits of your labours?

AH: That's a struggle. Building is very satisfying. It's also very meditative, for me, very clarifying. Someone could come in and say, "*This house is a piece of junk,*" and I can look them dead in the eye and say, "*You don't know what you're talking about.*"

But if he takes one of my books and says, "*This is a piece of junk*", I would have to ask, "*Why do you say that?*"

There's a central part of *Builder's Apprentice*, though, which does connect with what I do as a teacher. The book is about building, my experiences of going from being a novice to being a superintendent on a 29,000 square-foot house. But its central theme is the idea of following your calling, following your passion.

When I quit as an engineer, turned down Harvard and Berkeley, and had to sit down with my parents to say, "*I'm going to be a carpenter,*" they thought I was out of my mind. Everyone did, in fact. But I was doing what I was really excited about doing.

I think kids today should really hear that message. Because I think there's something dangerous in our society when we're teaching kids to start building their résumés from the 9th grade.

What we're telling kids is, *The measure of your life is what someone else sees in it on a piece of paper.*

That's a terrible message to teach someone, because at the end of the day, it's your life. And you only get one. Are you going to make the best contribution you can?

That really does drive a big part of my teaching. I challenge students by saying that the wrong question is to ask, *"What do you want to be when you grow up?"* Instead we should ask kids, *"What were you **meant** to be?"* That's a totally different question, a totally different thought process.

That is a theme in *Builder's Apprentice*: go find what that is. And even if well-meaning people tell you that you shouldn't be doing it, do it anyway. Do what really makes you excited.

HB: That's a perfect point to end on. Thank you very much Andy. It's been a pleasure.

AH: Thank you, Howard.

Questions for Discussion:

1. Do you agree with Andy that our society generally puts too much emphasis on "building a résumé" from an early age? Has this pressure increased, decreased, or stayed roughly the same over the past 20 years?

*2. How does Andy's belief in the importance of young people developing and following their passions fit with some of the central themes of **Flourishing** described earlier?*

Continuing the Conversation

Readers are encouraged to read Andy Hoffman's book, *Flourishing: A Frank Conversation About Sustainability*, that formed the basis of this book and entails the conversation between Andy and John Ehrenfeld on how to create a sustainable world.

And you can simply turn the page to read Part 2.

Saving the World at Business School

Part 2

A second conversation with Andy Hoffman

Introduction

Exceptional Times

I have never done a "follow-up" Ideas Roadshow conversation before, simply because the very concept didn't fit with what we are doing. I have always maintained that our mission is not journalistic, being much less focused on investigating, *What are you doing now?* and much more on exploring, *What led you to become what you are?"*

But what happens when "what you are" is inextricably linked to the rapidly changing political climate of our day?

Andy Hoffman is a business professor at the University of Michigan. He is also an environmentalist and a writer. Before all of that he was a carpenter and a builder. But perhaps the best way to define him is a passionate advocate for desperately needed social and cultural change.

His latest book, *The Engaged Scholar*, is a clarion call to arms to academics everywhere, urging them to take a much greater role in their societies.

> *"We produce our academic publications that go into our academic journals. We talk in our narrow academic communities and we move on to the next paper. We are paid a nice salary. We live a very privileged lifestyle. I think with that privilege comes responsibility. So, this is a call for academics to take on the responsibility that comes with the power and the place that they have in the academy."*

Well, you might be forgiven for thinking, *we've heard that sort of thing before*—this is hardly the first time a professor has urged his colleagues to move out of the ivory tower and take on greater social responsibility. And you'd most certainly be right. But two things make Professor Hoffman's case different from the usual sort of academic invocation to the public forum:

In the first place, he makes a pointed, and compelling, argument about the very structure of the academy, methodically taking it to task for its inflexible incentive structure that consistently reinforces academic isolationism and consequently minimizes its ability for societal engagement.

"The rewards here in a business school and other professional schools are primarily about academic publications. We do annual report review at the end of the year: research with a capital "R", teaching and service with a small "t" and a small "s". The journals that we consider the top journals within business schools—those that people would get tenure for—I'm sure most people in business have never heard of, much less read. And bringing your work into the general public is not really rewarded; people do it because they want to. And that has to change.

"It's not only unrewarded in terms of formal rewards like tenure, it's blocked by the informal culture as well. There's something called "the Sagan effect", named after Carl Sagan—the idea that if you bring your work to the general public, you're a "popularizer", not a serious researcher. Stephen Jay Gould, the famous anthropologist from Harvard ,was very, very aggressively opposed to that idea. He strongly believed that people who bring their work to the public are very important to the academy. But the problem is that right now the academy doesn't reward them for that.

*"So in **The Engaged Scholar**, I'm trying to encourage people to envision a scenario where individual scholars can say, "I want to spend a substantial fraction of my time as an academic in communicating ideas to the public" and be rewarded for doing it by their institutions, universities that strongly support basic research while also broadening the tent for the multiple roles that academics can play in elevating the role of the university within society."*

All of this will likely strike you as both appropriate and entirely uncontroversial—after all, who *wouldn't* argue for universities to play a more constructive role in our society? And who better qualified than a business professor to point a critical finger at the negative repercussions of inappropriate job incentives?

Why, you might be tempted to ask, *should I care about this **now**?*

Well, that's the second point, and the key one. You should care about this now because we are presently mired in a crisis—and I don't mean just the global pandemic and severe social and economic fall-out from that, which would certainly be bad enough—but a crisis of *understanding*, which has the potential to wreak vastly more harm still, long after the pandemic is a distant memory.

"A real problem we have now is that we've lost the ability to have discussions with people that are respectful and divergent. Somewhere along the line, we've developed this idea as a society that for me to be right, you have to be wrong. And that's simply not the case.

"The explosion of social media and the ability of people to create filter bubbles around themselves so they get information that confirms their worldview and biases plays a significant role—not just social media, but cable news too. There are all kinds of news outlets that are very tribalistic, just focused on getting people to absorb the particular flavour of information they are offering.

"Many people, for example, still don't believe that COVID is real. I was recently talking to a doctor at the medical system here at the University of Michigan who is treating a patient for COVID who didn't believe that COVID was real. Now, think about that for a second.

"You have people who don't believe that climate change is real. You have all kinds of wild conspiracy theories running wild right now, because people are getting a very filtered view of information that reinforces their worldview. And let's face it, there are people who are making a lot of money by feeding this situation."

Given this turgid, manipulative, bipolar, "fake news" morass so many of us are mired in, who better than the university to help us gain some much-needed perspective and genuine understanding.

> *"As purveyors of knowledge, the university can play a much stronger role in communicating that knowledge to the regular public and politicians. Of course there's going to be some blowback:* **You're just a bunch of liberal, latte-drinking professors!**

> *"Well, if that's the worst you can throw out at me, I think I can weather that storm in order to bring important work into the general public to serve our society and to serve the world. Because the dominant influence the United States has on the world can't be ignored. And the idea of balkanization is something we can't afford either: we're at 7.5 billion people, soon to be 10 billion people."*

Sometimes, the times you live in force your hand: the question is no longer whether or not universities are better off being ivory towers. What is at stake now is simply a matter of survival.

The Conversation

I. Reprise

7 years on

HB: I thought I'd begin with an anecdote that I received from my former video editor after our first conversation seven years ago. He said to me, "*I really liked the one with Hoffman, because there were all these sparks flying, there was lots of tension between the two of you.*"

And I remember having two bemused reactions to this. The first one was, *That's not what we're all about—Ideas Roadshow isn't supposed to be some deliberately controversial political talk show*. And the second was, *What on earth is he talking about?* That wasn't how I remembered our conversation going at all. I just remember having a good time talking with you.

AH: I'm the same way. I don't remember anything combative, but I love the banter back and forth. That's the way I teach classes too.

HB: Well, you see, you're a very reasonable guy. I guess I should start from the beginning by saying that, as it happens, I broadly concur with a lot of your core messages. And I think the phrase "as it happens" is worth focusing on, because—and this alludes to some of the things that you've written about—I think it's extremely important to be engaged in an open and candid dialogue with people, both those you agree with and those you don't.

So, the fact that there are many areas of overlap where we share common views is nice, but certainly not necessary or essential. And there are, of course, areas where we diverge, certainly in particulars, such as the fact that I think you have a tendency to err on the side of idealism, however admirable that might be. Personally, I have a little bit more of a sceptical attitude. But once again, respectfully highlighting the differences in how we view things makes up an essential part of what it is that Ideas Roadshow is all about, as you've even mentioned yourself in some of your work.

Another point that I think is worth highlighting straight off, almost as a preemptive strike, in case I get accused by someone else of being combative again—

AH: I don't think you're combative, by the way.

HB: No, no, I'm not talking about you. I'm talking about other people, like my former video editor. I mean, it's already been clearly established that you're a right-thinking, reasonable person.

But what I wanted to say is that the issues that we're talking about—business and societal change and climate change—these are issues that many people are quite understandably very passionate about. They are arguably existential issues for humanity. In other words, if you're not going to have strong, passionate feelings about this sort of thing, you could well make the argument that you don't have a pulse.

AH: I would add to this that a real problem we have now is that we've lost the ability to have discussions with people that are respectful and divergent. Somewhere along the line, we've developed this idea as a society that for *me* to be right, *you* have to be wrong. And that's simply not the case.

HB: Absolutely. And I know that you've written extensively about this, as I expect we'll get to this in more detail shortly. But I suppose that for my part I just want to emphasize that it's not just that we're living in an age when the medium of exchange of ideas and listening to other people with different views is overlooked and underappreciated, it's that I believe that doing so is particularly important in this moment in human history.

In other words, while I would always agree that respectfully treating the views of others is both polite and pedagogically important in our quest for knowledge—you mention encouraging banter in your classrooms and so forth—but it's even more than that, I think: when the stakes are as high as they are now with some of the issues that we're facing. In short, I would argue that proceeding in this way is nothing less than essential.

AH: Yes, I would agree.

HB: I think it's also worth explicitly highlighting that this is the only time we've done a follow-up Ideas Roadshow conversation, and I will naturally be periodically referencing that in terms of investigating questions such as: *What has changed in the meantime? Have things worked out the way that you had imagined they would when we first spoke? What has worked out well? What has not worked out well? How have your views evolved over time, compared to the way that they were seven years ago or perhaps earlier still?*

But in keeping with my earlier comments, it's perhaps worth emphasizing *why* I've decided to break Ideas Roadshow convention by having this follow-up conversation with you. Unlike one of our standard conversations on the subtleties of Renaissance history or dark matter, say, the research and scholarly issues that you concern yourself with are both particularly relevant to society at large *and* particularly time sensitive.

In other words, despite the fact that I'm always going on about how Ideas Roadshow is not journalism—and it isn't—there are nonetheless times when our mission *does* overlap with core aspects of the mandate of journalism, such as the importance of generating widespread understanding of pressing issues of direct relevance to all of us—and this is clearly one of these times.

So that's why I feel that this topic justifies a follow-up conversation, and I'm very grateful to you for your willingness to participate and give even more of your time to our project. So, thank you very much for that.

AH: It's my pleasure, Howard.

Questions for Discussion:

1. Do you agree with Andy when he says that, "We've lost the ability to have discussions with people that are respectful and divergent"? If so, why do you think this happened?

2. What do you think Howard means by stressing that "Ideas Roadshow isn't journalism"? What does it mean to be "journalism" these days and to what extent, if any, has that changed in the past few decades?

II. Truth Decay

Facts under fire?

HB: So while I'm merrily breaking convention, I'll do something else I never do which is to explicitly date this conversation by mentioning that we're having it in the midst of the coronavirus pandemic—in December of 2020—which is why, of course, we're doing it remotely.

Now I know that you have two books coming out in 2021, both of which you've graciously given me a peek at, so I'm naturally keen to talk explicitly about both of those. The first one I thought we'd talk about is called *The Engaged Scholar*.

Here's what I expected, based upon our previous conversation, *The Engaged Scholar* to be about. I expected it to be all about how academics have to go out into the social sphere, into the "real world" as it were, and play an increasingly large role in convincing people about the reality of vital threats like climate change and what can be done about it.

Well, there was some of that, of course, but that was not really the main thrust of what I read. In particular, many of the issues that leapt out at me were broader ones about the role of a university in contemporary society and the role of a public intellectual. Is that right? Is that wrong? What were your motivations in writing this book?

AH: Well, it's interesting. I agree with you everything you just said: I think that's what the book is about, and I see the two issues you mentioned as the same. The book is trying to make the case for a widespread recognition of a different role for the academic in society, one that is very engaged in a public and political discourse; and it's about the rewards of the academy standing in the way of that.

It's mostly a call to young scholars, to be honest with you, to define for themselves why they became professors in the first place. I'm willing to bet that, for 99% of them, it's because they wanted to have a positive impact on the world, not because they wanted to have a high citation count and a high h-index and other standard academic metrics.

I would also add that this is a book that I've wanted to get off my chest for a long time. But since we last spoke seven years ago, a lot has changed to make this book even more relevant and more important.

I hang the book, at the beginning, on a report from the RAND Corporation—which, interesting enough, is basically a defense department think tank. They came out with a report called *Truth Decay*, which had four conclusions.

One, we are debating facts.

Two, we blur facts and opinion to an alarming rate—it's important for all of us to recognize, when you see a news story, what's fact and what's opinion.

Three, we distrust previous trusted sources of information.

And four, all of this is happening at a level not seen in this country or in the world, in centuries. And social media is a big driver of that.

To me, all of that represents a call to academia to step into the fray, step out of the ivory tower and start to bring scholarly work into the general public. I think that we should do that for the sake of society, but I also think we should do that for the relevance of the academy, which is becoming increasingly irrelevant.

We produce our academic publications that go into our academic journals. We talk in our narrow academic communities and we move on to the next paper. We are paid a nice salary. We live a very privileged lifestyle. And I think with that privilege comes responsibility. So, this is a call for academics to take on that responsibility that comes with the power and the place that they have in the academy.

HB: You said that you've been thinking about these ideas for a long time, and you also mentioned that things are in a much more perilous state now as compared to seven years ago. Why do you think that is?

What's happened in the intervening seven years that have caused things to become so much worse now than they were then?

AH: Well, certainly the explosion of social media and people's ability to create filter bubbles around themselves so they get information that confirms their worldview and biases plays a significant role—not just social media, but cable news too. There are all kinds of news outlets that are very tribalistic, just focused on getting people to absorb the particular flavour of information they are offering.

Many people, for example, still don't believe that COVID is real. I was recently talking to a doctor at the medical system here at the University of Michigan who is treating a *patient* for COVID who didn't believe that COVID was real. Now, think about that for a second.

You have people who don't believe that climate change is real. You have all kinds of wild conspiracy theories running wild right now, because people are getting a very filtered view of information that reinforces their worldview. And let's face it, there are people who are making a lot of money by feeding this situation.

HB: Absolutely.

AH: We don't know who to trust anymore, as a society. We're divided on who we trust and that is quite dangerous.

HB: I won't argue with you about what is happening on the ground—the significantly increasing number of people who believe crazy, erroneous things—because it clearly seems to be borne out by the facts. But I'm nonetheless very confused by it, because I don't understand it at all. You say, "*We don't know who to trust anymore*"—that's a statement that I hear a lot these days. But it actually makes no sense to me.

Personally speaking, I don't have any more difficulty today than I did in 2013 to get an understanding of where reputable news is coming from, where corroborated, independent, scientific facts are coming from.

Let me give you one specific example out of many. I have a subscription to *The Economist*, which in my judgement to all intents

and purposes hasn't substantially changed at all in the past 7 years—
or in my view since I started reading it 30 years ago, or perhaps since
it began publication in the 19th century.

And in fact, an irony that many people have pointed out is that,
if you are of sufficiently reasonable curiosity and determination, you
can certainly get more information today than you could have 20
years ago.

For example, as the pandemic developed, I've learned all sorts
of things about different types of vaccines and different aspects of
biochemistry and biotechnology, none of which I knew anything
about a year ago. This information is freely available from a number
of different websites, whether it's the CDC or the FDA in the US or
the European Centre for Disease Prevention and Control or whatever.

So personally, I hardly feel less informed or more tribalistic
or more driven to conspiracy theories now than before—quite the
contrary, in fact: I feel that I have *more* opportunity to be able to get
objective, unbiased, reasonable, accessible information now than
I've ever done.

So frankly, I'm mystified by all of this. I mean, it's clearly true—
you're clearly right: things *are* vastly more tribal and conspiracy
theories do now abound like never before. It seems like a substantial proportion of people in the Republican Party these days believe
in, or at least tolerate, QAnon or some damn thing. I mean, there's
obviously all this nonsense that's going on, but **why**? What on earth
is it that is causing people to feel motivated to believe such idiocy?

AH: Well, what you're offering is a challenging question because I am
the same as you in the sense that I still read *The Economist*, *The New
York Times*, *The Wall Street Journal* and others; and I will also force
myself to read and watch other sources, like Fox News, to see what
kind of a diet other people are getting. It's important to recognize
the totally different worldviews that are presented in these different
media outlets, but sometimes it's a personal struggle.

The dream of social media was exactly as you described it:
democratizing knowledge, enabling people to get knowledge very

easily. That's the dream within the university system: that we can create online education that can be reaching people all over the world with greater ease, open access courses and so forth. The dream is also a global common, where people can come and debate ideas.

But it's had a darker side to it as well. Every technology is going to do that: a technology comes in and then the cultural ramifications start to play out and we have to start making corrections. For social media, these corrections involve moving from simply allowing people to get information to being somehow able to evaluate the credibility of the information. How we're going to get there, I'm not exactly sure, but that is the great challenge of the day.

In that RAND Corporation report I mentioned earlier, they conclude by saying, "*This is the existential crisis of our time.*"

If we can't agree on facts, how can we actually address issues like climate change?

HB: Well, how can we address *any* issue without an understanding and acknowledgement of the facts?

AH: That's right. So we have to ask ourselves what kind of private interests are coming into this domain to shift the conversation, like certain economic interests who want to weaken the veracity of the science of climate change because of the threats to their economic interests. There is a whole litany of pseudo-scientific journals. Certain kinds of think tanks have a very political agenda and are not afraid to stretch the truth or present things as if they have academic rigour when they don't.

I struggle. I also always try to question whether or not there is bias in the information that I get that I previously trusted. For example, *The Economist* will typically take as unquestioned the idea of global trade as always being good...

HB: OK, but having a philosophical bias—which you're absolutely right about, they certainly do; and they often deliver it with an annoying tub-thumping regularity that can certainly grate and one can often disagree with in the particulars—is not at all the same as distorting *facts*. I mean, I don't work for *The Economist*, but I would certainly recommend them to anybody as *one* important, trustworthy, source of information. But my main reason to mention them in our discussion was that I don't think they've substantially changed one bit in the past 7 years, so logically they can't be associated with this mysterious rise in "fake news" or "untrustworthiness of the media" or whatever you want to call it.

AH: I would say, for example, that *The New York Times has* changed. Their coverage has really become a counterweight to Trump, in my opinion—they really have taken that position. And MSNBC has shifted much further left—as sort of a counterbalance, perhaps, to Fox News.

So the landscape has shifted and we always have to evaluate, *What's the quality of the information and what's the bias within that information?* Ideally, we want something in the middle, unbiased. And on the top, if we were to draw a matrix, we want the most rigorous, most balanced.

There is another news magazine out there which bears mentioning called *The Week*. I really like their format. They give a short tidbit of factual information, and then give the editorial opinions of that information across the spectrum. It makes for very interesting reading.

Questions for Discussion:

1. Do you think that there is more "fake news" out there today than there was 10 years ago? If so, why do you think that is? If not, why do you think that many people believe there is?

2. To what extent does the very structure of a corporate news agency naturally inhibit objectivity?

3. Which news source would you regard as "the most objective" and why?

III. The Value of Wisdom

Beyond knowledge

AH: I would also add that, in the academy there's a certain bias that scientifically derived knowledge is always best. There's something called the Data-Information-Knowledge-Wisdom pyramid. Picture a pyramid. On the bottom we have all this *data*, which we then turn into *information* by recognizing patterns or connections; and we then turn that information into *knowledge*, factual knowledge, by running regression analysis and things like that.

But the next step is turning that knowledge into *wisdom*, which requires some set of aspirational principles that says, *These are the values we stand by.*

And in my opinion, the academy is not very good at creating wisdom. It's very good at creating knowledge. Yet I'm surrounded by a lot of people who think that what we produce in our academic journals is the final say, and communicating that with the public is not our job. I disagree with that on several levels.

First, is it always the final say? An agronomist may know a lot about soil science and doing plantings and things like that, but the farmer who's been cultivating his or her fields for the past three or four decades has a great deal of tacit and pragmatic knowledge that *also* has value, yet most scientists would dismiss that value as irrelevant.

It would be like talking to a pianist. I, as an academic can study how a pianist taps the keys. Do I know how to play the piano? Absolutely not. How do I get that knowledge out of the pianist? Well, it's not scientifically derived and we don't know how to capture that within the academy. Maybe I'm getting a bit too esoteric here.

HB: No, I don't think so.

AH: I'm trying to capture the idea that we should *all* analyze what kind of information we rely upon and ask ourselves, *Does that blind us to certain things out there that we should be aware of?*

And that speaks to the idea that you started off talking about, of being able to talk to each other. Because for an engaged scholar to successfully enter the public and political discourse, they not only have to leave the ivory tower and bring their work to the public, they also have to *listen* to the public. And they also have to listen to politicians and others so as to be able to best temper, balance and present their work in a way that's accessible and respectful of other kinds of knowledge that's out there.

HB: Well, you've touched on a lot of points. First off, I suppose it's worth making a clear distinction between our two situations. As a university professor—and a business professor to boot—at a major American university who is passionately committed to encouraging your fellow academics to play an increasingly impactful role in the public space, you are both more involved and more constrained than I am.

When I hear something nonsensical in the media—which is hardly a rare phenomenon, I must say—I can simply say to myself (and often do), "*These are crazy people saying crazy things*," and then turn off the TV or computer or whatever and open up an interesting book instead. It's different for you, of course, because there's a real conflict here with your personal values and perceived responsibilities that you discuss in *The Engaged Scholar*, which I'd like to get back to later.

But there's another point, which also came up in our previous conversation, that I think is really essential. You talk about this pyramid with data, information, knowledge and wisdom, and what wisdom means in terms of connecting with our values and even fully understanding and appreciating what those values are and should be. That seems very compelling to me. But I think it's also worth emphasizing something else that you also alluded to, which is that there are many times when, just because you have a lot of data and the right amount of scientific rigour, that doesn't guarantee that your theoretical interpretation or understanding is going to be correct.

As you know, I come from a scientific background and I've seen lots of people use and misuse data in all sorts of ways, or come up with completely plausible, theoretical explanations for a phenomenon that turned out to be completely wrong. So it's important to remember, too, that just because you're a scientist, or just because you are associated with other people who are scientists, that hardly gives you privileged access to the truth all the time.

Another issue you touch on which I think is worth highlighting is that sometimes scientific ideas are expressed in a very pedantic and condescending way towards members of the public, which naturally encourages a sort of backlash: *Oh, you little peons. You won't be able to understand this, so just trust us because we are the great minds.*

Now, I should say that I ***don't*** believe that most official scientific bodies or most scientists generally act in this way. In fact, I believe that most of them are very much in favour of an open transmission of knowledge to the general public and don't do it in a condescending way at all, but it's also undeniable that some unfortunately do.

And I think it's very important to be sensitive to that, to be very aware—as you've said yourself—that it's not just a specific message in and of itself: it's how that message is communicated and the corresponding dialogue that ensues with others. It's important to appreciate that if someone is a farmer in Alabama he might have a different background and skill set from yours that might make it particularly challenging for him to decide, reasonably and objectively, what's right, what's wrong and which path to follow on specific scientific issues when he's being deluged with all sorts of information from all sides. It's important to be able to look at it from his perspective.

AH: There's a number of things you're triggering in me. I think that one thing that's happening right now in our society is that we are looking for convenient labels to dismiss the other side. David Brooks of *The New York Times* once said, and I really love this quote, "*We don't debate people's ideas. We question their motives.*"

There are, indeed, people who say, "*I don't trust scientists.*" Well, that's not entirely new. The book that won the Pulitzer Prize in 1964 by Richard Hofstadter, *Anti-Intellectualism in American Life*, described this. Many people just don't trust scientists.

But in today's world, what we'll find is that people will look for a ridiculously extreme example and then paint the entire group with that brush. Two years ago, there was a survey that found that 58% of registered Republican voters in the United States felt that universities had a net negative influence on society.

And what I think they're reacting to is something that scientific institutions are also warning against—a phenomenon called *scientism*: the near religious belief that science, particularly the physical sciences, holds all the answers.

Let me give you what I think is a revealing example from personal experience. I once wrote a piece for *The Conversation*. What I was trying to say in the piece was that science has an important place but sometimes it should understand its limitations and practice a little humility. I anchored it on a line by Aldo Leopold that I've always loved. He said, "*In the final analysis, appreciation of nature boils down to an act of intellectual humility.*" So, as we start to entertain ideas like geoengineering the planet, maybe we should pause and ask ourselves, "*Do we really know everything?*"

Well, I got a rebuttal from a scientist. I'll refrain from using any identifiers, but this scientist went at me very aggressively. In the article, I wrote, "*There are certain things that science can't explain. One of them is that it can't explain love.*" And in his rebuttal, the scientist came back and said, "*Oh, science can most certainly explain love. It's merely the release of certain chemicals in your body. If we can inject these chemicals, we can elicit love.*"

Now, scientifically, maybe I can show that when you are in love, you are releasing certain chemicals. But if that is love, that represents a worldview that you are merely a bag of chemicals, that there's nothing in there besides the reaction of certain chemicals—and that's a worldview that I reject.

HB: Well, that's because you're a logical thinker and know the difference between causation and correlation. You don't have to tell me. For the better part of a decade I ran a scientific research institute, so I'm very well acquainted with the dogmatism and philosophical bankruptcy of many scientists. I could tell you stories, but that would take us much too far afield, I fear.

AH: Roger Pielke has written extensively about the role of science in policymaking, and one point he makes very strongly is that scientists should put their work out into the public, but then they have to step back and let the public and the political process work itself out. They don't have the final say. A lot of scientists bristle at that. They say, "*My science says X and therefore people should do Y.*" That's not necessarily true.

Look at the pandemic right now. Science says we should all go into a closet, close the door and come out a year later. Obviously, that's not realistic, both socially and economically. We have to bring in the pragmatic realities of being a human being in a social world. How do we do that? How do we thread that needle? How do we make that balance? That's something that data isn't necessarily going to be able to help us with.

HB: Right. And I think humility is very important, because you have to recognize your appropriate domain. You may be an immunologist, you may be a virologist, but that doesn't mean that you're necessarily in a position to look at the bigger picture in terms of people's mental health, wider economic factors, and so forth. I mean, we all know this, right? If people are going to commit suicide because their business has collapsed, it doesn't make too much difference what the molecular state of their body is at that particular moment.

AH: Right.

Questions for Discussion:

1. Do you feel that most scientists are sufficiently understanding of the views and perspectives of non-scientists?

2. Why do you think Howard mentions "the difference between causation and correlation" here? What, exactly, is he referring to?

3. How might it be argued that intellectual humility is not only vital for wisdom, but is also necessary for the development of scientific knowledge itself?

4. How do you think most scholars would react to Andy's categorization of "wisdom" is distinct from "knowledge"?

IV. Investigating Rewards

Intrinsic vs extrinsic

HB: I'd like to talk some more about this issue of scientism and scientific hubris, but first I'd like to back up and specifically address some of the more general comments you make about the academy in *The Engaged Scholar,* because my reading of the situation is a bit different than yours. So let me give my perspective on things that you can react to as you see fit.

I think there's a bit of an amalgamation going on when you talk about "the academy", with my perception of the university environment being somewhat different: I look at research, broadly, in terms of three different classes or types.

In one class we have people working away on fundamental basic research, purely curiosity-driven things, whether you're a number theorist, a classicist, or somebody who's comparing and contrasting different perspectives of medieval chronicles. These are people who are motivated to do research for research's sake, driven by pure personal and intellectual interest.

Then you have people who are involved in what I would call more applied research, such as someone who is working in material science say, exploring different aspects of vulcanization or rubber or something like that; or a geneticist who's examining different gene sequencing algorithms or procedures. The idea here is that the specific end might be somewhat nebulous, but there's an obvious overall applied thrust to things.

And then there are those who are tackling issues that they believe to be pressing societal problems: those who are investigating climate change or contemplating how we can reduce wealth inequality in our society, say. Now different people would naturally agree or disagree with their characterization of whether or not something is "a pressing societal problem", but that doesn't matter for my classification scheme. The only thing that matters is how *they* would define what they're doing.

And the point is that if you're in that third category, as I interpret you and many other people to be, then not only is it *significant* or *relevant* for you to be thinking about how to communicate your ideas to the general public, government agencies, business community and all of that, but I would say that it is actually *essential*. Because the idea of sitting there and saying, *"Well, there's this existential threat to humanity, and I'm just going to write a paper and worry about my citation rates,"* is inherently ludicrous—the very nature of the problem necessitates that you have to get in there and try to somehow publicly address it.

So that's why I was a little bit surprised that you were looking at the role of universities generally. For me, the question of the role of universities is more complicated given these different classes of researchers and scholars I was just talking about; I would say that what we're really talking about here is the last class I mentioned. But perhaps you would disagree with that.

AH: Well, you're presenting a model that I mostly agree with. I almost think about it like a supply chain: you have people doing a spectrum of both basic and applied research. But even that is a framing that's too simplistic—I'm not so sure that it's true, this idea that there's a bifurcation between those who do basic research and those who do applied research.

That's a remnant—I get into some of the details of this in *The Engaged Scholar*—of the famous report by Vannevar Bush, the former dean of engineering at MIT and head of the Office of Scientific Research and Development during World War II, to shape post-war research in the United States. And many people now are starting to question this sense of a bifurcation, believing that these two things actually start to overlap.

There's a famous book by Donald Stokes called *Pasteur's Quadrant*, describing how Pasteur did research that was rigorous, basic, *and* applied at great importance for society. And Roger Pielke, in his book, *The Honest Broker*, says that first class that you described—people doing research just for the sake of knowledge, without any awareness of the world around them or the politics of it—that's a fiction that belongs in Hollywood movies that doesn't really exist.

HB: Hmmm. Well, I've seen it.

AH: OK, leaving that aside, the second class that you described are those who take their work to the world. You might imagine, for example, professional schools like the school I'm in—a business school—would fit into that category. And I'm here to tell you that the rewards within professional schools hearken back to that basic model. The rewards here in a business school and other professional schools are primarily about academic publications. We do annual report review at the end of the year: research with a capital "R", teaching and service with a small "t" and a small "s".

I'm not the only one speaking about trying to bring practice and engagement into the academic reward structures. The journals that we consider the top journals within business schools—those that people would get tenure for—I'm sure most people in business have never heard of, much less read. And bringing your work into the general public is not really rewarded; people do it because they want to. And that has to change.

It's not only unrewarded in terms of formal rewards like tenure, it's blocked by the informal culture as well. There's something called "the Sagan effect", named after Carl Sagan—the idea that if you bring your work to the general public, you're a "popularizer", not a serious researcher. Stephen Jay Gould, the famous anthropologist from Harvard, was very aggressively opposed to that idea. He strongly believed that people who bring their work to the public are very important to the academy. But the problem is that right now the academy doesn't reward them for that.

So, in *The Engaged Scholar* I'm trying to encourage people to envision a scenario where individual scholars can say, "*I want to spend a substantial fraction of my time as an academic in communicating ideas to the public*," and be rewarded for doing it by their institutions, universities that strongly support basic research while also broadening the tent for the multiple roles that academics can play in elevating the role of the university within society.

I gave a talk last year to a bunch of social scientists who were interested in sustainability and climate change. And at the end of the talk, I said, "*Okay, I'm going to do a survey right now. Everyone raise your hand if you care about the issue of climate change.*" All the hands went up. Then I said, "*Now keep your hand up if you devote the majority of your time towards academic publications.*" They stayed up. And then I asked, "*How many of you think that your next academic publication is going to do anything about the issue of climate change?*" And most of the hands came down. And I said, "*Does anyone see a problem here?*"

That's what I struggle with; and it's something I've struggled with since I became a professor. I didn't get into this just to formally participate in the academic literature. Of course I recognized that I needed to do that to get a job at a great university like the University of Michigan. But from this platform, especially with the position of full professor, I believe that I not only have the opportunity, but also the *responsibility,* to recast my idea of what my job is as a professor in order to increase its contribution to society.

Jane Lubchenco, when she was the president of the American Association for the Advancement of Science, wrote a paper about what she called "Science's Social Contract" arguing for a greater sense of social responsibility in the scientific community: We do all this work. We are funded in many cases by the taxpayer or by tuition paying parents or tuition paying students. And what is our responsibility that comes with that contract? Our responsibility is to use that work to try and contribute to society and not just inflate our citation count and our number of A-level publications, which is the coin of the realm for the building I'm sitting in right now.

HB: Well, I've heard that sort of thing quite often, and it invariably leaves me feeling vaguely irritated. Let me try to explain what I mean. Academics often say things like, *"I'm funded by the taxpayer. I have a moral responsibility to go out there and communicate my work to the taxpayer."*

So on the one hand, of course, you want to have transparency; and clearly you shouldn't be funded by the taxpayer and then do something against the taxpayer's interest. And clearly, too, you shouldn't be in such a situation where you're concealing what it is that you're actually doing—unless you're funded by the department of defense or something, which is a whole different matter entirely.

But whenever I hear people invoking these sorts of arguments, it touches on something else that you write about that I feel is very important, which is the importance of having a real sense of passion for what you're doing—that is, your personal values.

Let me try to be more specific. Andy Hoffman, I'm guessing, doesn't feel motivated to alert people to the dangers of climate change and how we might best address it simply because he says to himself, *"Well, I'm funded by the taxpayer, so I should really go out and do that."* He feels motivated to do it because that's an essential aspect of who he *is*—that's what he believes in.

In other words, I think what this really boils down to is that we should have people who are in academe for what I would simply call "the right reasons". To me, it's simply a matter of values. If you decide to become an academic because you want to have a huge citation rate, or because you want to be considered a big star by other people, or because you want to be a dean or whatever, then quite frankly you bloody well shouldn't be there in the first place.

Call me a raging idealist, but I believe that you should really only be there if you're in it for the right reasons, which should simply be a reflection of your passion and values, and hopefully—that would be nice, but I'm not convinced it's essential, actually—your motivation to positively influence society.

AH: I would interpret what you're describing is highlighting the difference between my implicit rewards—*Why am I doing this?*—and the explicit rewards—*What are people telling me to do?*

And my point is that the explicit rewards drive people in a certain direction. When you're going for tenure, all you're focused on is the number of A-level publications. Period. That's what the system does to you. And over time, people get indoctrinated into this system. The preface to *The Engaged Scholar* opens with a specific question: ***Why did you choose to become a professor?*** *Reflect on that. Think about it carefully. Because if you don't know exactly why you're doing this, you will become what the system wants you to be.*

Herbert Shepard, an organizational scholar I really admire, wrote a famous essay called "On realization of human potential: A path with a heart". In it he warns people, *"Don't become a cormorant."* A cormorant is a bird that's really good at catching fish. So the fisherman puts a band around his neck and a rope around his foot and throws it into the water. And all day long, the bird comes back and the fisherman just takes the fish away from him before sending him back in the water. That is someone who is just following the extrinsic rewards. If you don't push back and say, *"This is who I am, and this is why I chose to do this,"* you will fall prey.

I often talk to doctoral students who feel that in the course of their doctoral studies that passion has been squelched into a more mechanical pursuit of the attainment of the symbols of success within the academy. And I want to undo that in this book. I want to keep that passion alive, because at the end of the day those are the people that are going to have the most exciting careers.

It's an appeal to meaning and purpose—that's really the heart of the book: *What meaning do you get out of your work? What purpose do you find from it?*

Whenever I hear someone say, *"I'm not going to do that because I won't get enough rewards for it"*—well, I've got to be careful what I say here, but what a sad statement: feeling forced to do what the rewards tell you to do. That's a very externally-driven life.

HB: To push the cormorant analogy a little bit further, you can also imagine people lying on their deathbed assessing their lives, wondering, *Have I just been doing something so that somebody else was getting fish all day long? Is this how I spent my life?*

AH· Yes I co-wrote a really enjoyable paper with A.R. Elangovan looking at the different stages of an academic career. I've seen many senior emeritus professors who start to become a little embittered because the world never fully recognized their genius.

And my response to that is, *"Well, you hid it under a bushel."* The idea that I should just write papers for academic journals and wait for the world to discover them and apply them is foolish. It's not going to happen. You've got to bring it out to the world. And if you don't do that, there's a good chance the world's never going to see your work. It's that simple.

So I agree with you entirely: at the end of my career, at the end of my life, when I look back on the things that I'm going to feel most proud of, it's not going to be my citation count on Google Scholar.

HB: At the end of our last conversation, you gave a very moving account of what you tell, not so much your doctoral students— although perhaps you tell it to them too—but your undergraduates. You say to them, *"Don't look at what other people tell you that you **should** be doing. Ask yourself what were you **meant** to be doing. Ask yourself what is it that can manifest your own values, your own beliefs, your own desires to live a flourishing life."*

Questions for Discussion:

1. Are you surprised at the notion that many academics need to explicitly ask themselves why they became a professor in the first place? Do you think this applies to some areas of scholarship more than others? Are academics more prone to blurring the distinction between intrinsic and extrinsic rewards than others? Less prone?

2. Should all scholars be strongly encouraged to engage with the general public about their work?

3. Do you think that researchers who are motivated to pursue "knowledge solely for knowledge's sake" actually exist?

V. Concrete Opportunities

And concrete scepticism

HB: OK, so I clearly believe in your core arguments—*"you had me at hello"* to quote the immortal *Jerry Maguire*. But what can we actually *do*? How can we realistically get to a situation where universities will shift even a little bit in recognizing that we shouldn't always use the same old metrics or the same old values, or the same old status symbols, or what have you.

And it's a complicated issue, of course, because as you know very well, I'm *not* suggesting that universities should simply do away with publications or peer-reviewed journals or any of that. I'm certainly not suggesting that.

But how can we move towards a broader, more human-centric, shall we say, view of what not only a scholar should be all about, but what the university should be all about in terms of improving its societal impact. How do we actually *do* that? I'm with you: I'd like to see that happen, but I don't see it going in that direction. And with all due respect, I'm not sure how many books you'd have to write before things *would* actually start to move in that direction—not just you, in fact, but everyone else writing books all over the world. So... how do we actually get there from here?

AH: Well, first of all, we have to think about it on an institutional level. And what I mean by that is that *all* the pieces have to move. You can't have just one school that decides, *"Okay, we're going to develop an entirely new reward structure."* A junior faculty member would have to be a little crazy to follow that unless they were guaranteed tenure—the entire system has to change, which includes the accreditation bodies. Well, we're starting to see some movement there. As far as the journals go, we're starting to see movement there too. Then there are recruiters and training. All across the board, all the pieces have to move. I do think I see positive movement here; I believe that changes are afoot.

HB: Okay, give me an example. Tell me something concrete. Give me something to hang on to.

AH: In my world, the business school world, the accrediting agency is the AACSB—the Association to Advance Collegiate Schools of Business. And the AACSB is starting to change their accreditation criteria to add impact: they want to measure impact to society in the business school to give them accreditation. There's a group called the Responsible Research in Business & Management network that's trying to get the top academic journals to start to focus more on academic papers that have relevance to the critical issues of our day—in particular focusing on sustainable development goals.

There are a number of training programs to teach young professors how to engage with the general public. I had the good fortune to be selected as a Fellow for a program like this at Stanford, the Leopold Leadership Program, and there are several others around the country as well, often with links to the AAAS and the National Academies of Sciences. They're looking to improve how scientists can communicate with the public.

In 2015 I ran a conference on this topic here at the University of Michigan. And to my amazement, during the opening panel discussion, I got *four* university presidents to sit on stage together for a moderated discussion on this topic: Arizona State, Virginia, Dartmouth and Michigan. Try to get *one* president on a stage for a discussion, that's challenging enough—I got four. So that's another sign, I think, that the time is right: people are starting to focus on this.

And then, just to add one more piece to this whole equation, the best way to change any system is to have a massive disruptive event. Well, we're in one right now: COVID is really shaking up the university system.

What we're doing right now, having this conversation remotely through technology, would have felt very uncomfortable a year ago. Now we're all used to it. And a lot of this is going to stick when we're done. The business model of the university is in flux—let's see how it plays out. But there are going to be a lot of changes after COVID, after the vaccine is distributed.

In the business-school world, before COVID, a lot of people were starting to seriously question capitalism, questioning to what extent it serves society. A statistic I often refer to is that before COVID 40% of Americans could not pull together $400 in emergency funds.

Think about that for a second. We're *in* that emergency right now. And come December 31st, if Congress doesn't act, a lot of people are going to be put on the street in the winter. This is a very scary situation. At the same time, by 2026 Jeff Bezos is poised to be the world's first trillionaire. And the stock market is booming.

Anyone looking at those diverging worlds who can say, "*The system is working properly,*" is not looking with clear eyes. And that was *before* COVID. Now COVID is here, and the conversation is becoming even louder.

The World Economic Forum has entered the fray and said, "*Okay, let's throw out this idea that the purpose of the corporation is just to make money for shareholders.*" That's a ridiculous idea, it doesn't work anymore: the shareholders are doing well, but record numbers of people are unemployed. Something is really wrong here. And there are numerous others, like BlackRock or the Business Roundtable coming out and saying, "*We need to rethink the purpose of businesses and society.*"

HB: Well, you had me right until you started talking about The World Economic Forum and the rest of that cabal. Once you start quoting those guys it all goes south for me.

AH: Well, those are aspirational statements.

HB: It's pretty clear to me that the biggest aspirations those guys in Davos have is to continually make ponderous, self-aggrandizing pronouncements that promote their interests and give them a platform to come across as the world's greatest geniuses. So I appreciate that you live in that world and have to be careful. But I don't.

AH: Well, we need both. We need critics like you to say, *"That's all nonsense,"* and we need people on the other side saying, *"OK, you **said** that; now let's start to engage in the debate and take it seriously and answer how we're going to implement what you promised."* We need both. We need the likes of you and Anand Giridharadas with his book, *Winners Take All*, to be throwing rocks. And we need people on the inside saying, *"Look at those people throwing rocks that made you say what you said—what do you mean by that, how are we going to go forward, and what role can academia play when we do?"*

HB: As you were speaking just now I suddenly thought of this very memorable scene in *Inside Job*. There are many great scenes in *Inside Job*, but there's this particularly revealing one that sticks out in my mind when Charles Ferguson is talking to Dominique Strauss-Khan who was then head of the IMF. Of course we can talk more about Dominique Strauss-Khan in another context, but let's not. Anyway, he was talking about meeting with all these investment bankers and CEOs right in the depth of the crisis when they candidly told him, *"We're too greedy, you need to regulate us."* And that moment lasted—I don't know—maybe ten femtoseconds, or something like that. And then it was gone. And everything went right back to the status quo.

So when you tell me about the World Economic Forum, I think to myself, *Here we go again—this is another occasion when these guys will simply say whatever they feel that they need to say to get from one moment to the next relatively unscathed.* So I agree: *if* we have people who can actually hold their feet to the fire after that moment passes, *then* maybe there's an opportunity to actually go forwards. But I'm pretty sceptical.

AH: Well, that scepticism is important—it's very important. Joseph Stiglitz, a Nobel Laureate in Economics, said that capitalism *"needs to be saved from itself"*. Of course I, as a business school professor, have to recognize my domain of influence. I do try to impact the business world, but my biggest impact is through my students. We're going to keep graduating business students, and I want to change what kind of students we graduate; I want students with more of a conscience.

There was a great piece on *Medium* last year by James Gamble, who used to be a corporate attorney, called "The Most Important Problem in the World". He wrote, "*The system that we have compels business executives to act like sociopaths, to focus only on themselves and disregard the interests or the values of anybody else around them.*" In my domain of influence, I want to change that. I want to have graduates come out of the business school with a much stronger conscience over what kind of legacy they want to leave.

Just like the professor who comes to the end of her life and looks at her h-index and citation counts and is unsatisfied, I really think that someone coming to the end of his life and has a lot of money but left wreckage behind him is not going to be terribly satisfied either.

So I want to instill that in business students *now*. Which is why I believe that the idea of "management as a calling" is such an important aspect of the future of business education. It's one clear way how we can try to change one piece of that institutional fabric in order to drive change.

Questions for Discussion:

1. If Joseph Stiglitz is right that "capitalism needs to be saved from itself", who, exactly, do you believe is best placed to do the saving and how might it conceivably be accomplished?

2. Do you share Howard's scepticism, or do you believe that statements Andy highlight from the likes of the World Economic Forum are a sign of genuine social progress?

3. How do you think the COVID pandemic will impact global social and economic policy in the next 5-10 years?

VI. Management as a Calling

Focusing on the students

HB: So that's a great segue to your other book coming out soon, *Management as a Calling*. There were a couple of really interesting points in this book that I'd like to highlight. The first was a fairly obvious structural issue when you think about it, but something I had never thought about before in any detail.

You highlight how, while it's all very well and good to talk about the potential of harnessing student passion and imagine them leaving business school suffused with energy and starting groundbreaking new businesses that will change the world for the better, the very structure of the system actually *detours* innovation because their tuition rates are so high that by the time they leave business school they will have such a high debt load they will be far more likely to be lured into a safe, conservative status-quo job that will allow them to pay back their debt.

So there's a really large structural failing there that struck me as quite important. Do other people recognize this? Is this concern reasonably widespread? And, if so, what are the concrete plans to try to address that?

AH: Universities are vividly aware that the rising cost of tuition is a problem that has to be addressed, which is a principal driver behind many of the fundraising activities that universities are doing. For many public universities, like the University of Michigan, the amount of money we get from the state is actually incredibly low—I think it's only something like 14% or so—and the result is that we have to fundraise.

A big part of the fundraising effort goes towards the creation of an endowment to offset tuition costs. It has to be done. Michael Crow, the president of Arizona State University, has been very vocal about this, saying publicly that if we stay on our current track higher education will become a luxury for the rich. And once that happens, we are doomed. So we have to correct this, we have to fix this. This is on the agenda of every university president, the rising cost of tuition.

And it carries with it other perverse effects as well. Here in the business school. I see a lot of students that begin their education saying, "*I'm going to change the world. I don't care about money. I'm going to use my business education to make the world a better place.*" And then it comes time for graduation, and they start to look at the salaries of the positions they're most interested in. And then they look at the salaries in consulting, where the average salary is something like $150,000, with a $40,000 signing bonus. So it's natural that many will start to yield and say, "*I'd rather go for that.*"

The end result is that a lot of students who go through business schools are channelled towards consulting and finance. Whether they like it or not, that's where the money is, so that's where the status is: it makes them feel like they've gotten the most out of their education, which is a sentiment that is fed into them from the start.

There are many rankings of business schools, or schools in general, in terms of "return on investment", which I really find revolting. If "return on investment" is your metric for going into education then I really don't think you're going to live a life that you chose. You're going to live a life driven by money, which is a very sad way to live.

And this idea of conformity is something that we do to people from a very young age now. I sit on committees for high-school scholarships, and I'm appalled when I see students from the fifth grade spending hours and hours focusing on their résumés. We're teaching kids, *Your life can be summed up on a piece of paper to be evaluated by somebody else.* That's an awful lesson to teach people.

HB: Absolutely. The second point I had is related to that, I think, together with an anecdote from your book. You talk about how one young student—at least I presume she was a young student—confided in you that she felt that "her values were assaulted every time she walked into the building".

You didn't say it explicitly, but my guess is that she is precisely the sort of person whose values, by and large, would resonate with those you are trying to inculcate within your business school and the academy as a whole—it wasn't like she was a frustrated sociopath or something.

AH: Right.

HB: And that made me think of something slightly different, but also related, I think, to all of this talk of values—which is the testosterone-driven macho-strutting aspect of much of this sort of culture.

I'm guessing that, however difficult it might be for an idealistic man to survive in this sort of culture with his value system in tact—and I'm sure it is—it would be at least doubly difficult for a woman. Is there anything to that, you think?

AH: That's an interesting question, because I do think that the business curriculum is driven by many testosterone-laden measures. It's about "dominating the competition", "competitive advantage", and a lot of war metaphors. It's about "whoever has the most money wins" and "transactional relations".

I think that has to change, because there are places where cooperation is actually quite healthy and quite necessary, where the idea of recognizing connections within the social and natural environment is important.

So I think what you're describing is critically important. And I do think that there's a growing number of students who are coming in and seeing that for what it is. I taught a course this past winter that was joint between Michigan's Ross School of Business and Ford School of Public Policy. It was called *Business in Democracy, Advocacy, Lobbying, and the Public Interest.*

It was motivated by how stunned I am of how few business schools have courses on lobbying and fewer still on the idea of lobbying as a public service. And as the class filled up with students from both schools I asked them, *"What do your peers think about you doing this?"*

Students from the business school said that many of their peers were puzzled: *Why would you take a course on government? What does government have to do with business?* That is just mind-blowing, but it's the prevailing attitude within this building: *Government has no role in the market; regulation is an unwarranted intrusion in the market.* That's just nonsense. That's flat out nonsense. The market is designed and regulated by the government.

Meanwhile the public policy students said that many of their peers were disgusted that they would even step foot in a business school building, let alone take a course there.

So somehow we have to break this barrier down: we have to recognize the mutual importance of both. Public policy students often think that if government just sort of sets the rules, everything will be fine. And that's not true. Meanwhile, many business school students think that if you just get government out of the way, everything's going to be fine. And that's not true either.

And that animates many of the ridiculous debates that we have in this country, that it's either "more" or "less" government, not whether or not it's "the right kind" of government.

So, thankfully, I *do* think that there's a number of students coming into business schools now that want to use their business education to have a meaningful career and to make the world a better place, not just to make themselves rich. And we need to feed those students, we need to serve those students.

Quite frankly, if we don't, we're doomed. If we keep graduating business students that care about nothing except the size of their bank account, then we're doomed. It's that simple.

HB: Well we're *all* doomed. I mean, it's not just your business school, or business schools in general. The planet is doomed.

AH: That's right. But on the other hand, if we can turn around the idea of the purpose of the business in society, then we *can* solve the problems we face. The market is the most powerful orienting force on earth. You may lament that fact, but it's a fact. And while you might say that the market caused these problems in the first place—and that's most likely true, although the market is much bigger than just business and involves civil society, government, consumers, investors, the whole deal—the market *has* to solve these problems or they won't be solved. It's that simple. So if we don't put effort into changing the market, we are doomed.

HB: Well, I certainly agree that you have to pay attention to the market. Personally speaking, I'm hardly anti-market or anti-capitalism—I mean, I'm running a business after all, so being anti-market would be rather counterproductive, not to mention verging on the oxymoronic.

I've heard you say this sort of thing before on several occasions: that the market or business or capitalism or whatever is the most powerful force that exists. I'm naturally tempted to quibble with you about that sort of statement in a smart-aleck sort of way—what about things like dark energy or the force of the human spirit?—but I take your point: the market must be harnessed if we are to make progress.

So no real argument there. I would, however, like to pick up on what you said before about the culture of government versus regulation, because, frankly, that's very puzzling to me. I would have thought that most educated, reasonable people on planet earth are at least intellectually aware of the fact that markets have to be regulated, with the key question being, *To what extent* and *in what instances* should they be regulated?

So that's a bit confusing to me because I would have just assumed—naively, it seems—that any reasonable person is aware that there is a trade off: if you have too much regulation, you have too much red tape and not enough innovation, while if you have completely unregulated markets, you're not only faced with a sort of Darwinian inhumanism and rampagingly high levels of wealth inequality, you also naturally produce monopoly power that winds up gaming the system just as we saw so vividly in, once again, *Inside Job*.

Anyway, we don't have to go into that again, but it's worth emphasizing for the record that I find it very significant that you have consistently, and vocally, maintained that regulation to some degree is something which necessarily needs to exist.

But I'm still shocked by how this seems to make you an outlier, some sort of "extreme voice" within the professional business school community, so I'd like you to comment that: to what extent these views make you an outlier and if that seems to be changing at all. And secondly, another point I found quite interesting in *Management as a Calling* is something that you alluded to just now, which is that there are sometimes positive aspects to lobbying that often go unappreciated, so I'd like you to discuss that as well.

AH: Let me start with the second point first. I am not going to give you a Pollyanna view that a lot of lobbying was good. Last year there was $3.4 billion in lobbying in Washington, and on top of that you can throw in a presidential election with over $2 billion in PAC money. Clearly money has corrupted our political process. So the first thing I would do if I had a magic wand would be to start to pull money out of politics.

In my view a big opportunity to do that was just missed with the Trump administration. He came into office saying, "*I am not beholden to anybody and I'm going to drain the swamp.*" Had he stepped in and said, "*My first act as president will be to focus on campaign-finance reform; I'm going to get money out of politics,*" I would have sat up and said, "*Oh, this is interesting. Let's go with this. We need this.*" So, money in politics is a significant problem and I'm not going to say that it isn't.

If I could wave a magic wand, I would also make it totally transparent: *You want to do some lobbying, fine. But you have to disclose all of it, not just what you do, but the trade associations that you support, the PACs you support, everything you do is now going to become visible.*

And then I would institute a number of other controls to create more of a barrier.

But from there, things get much more complicated. You can go back to Adam Smith. Adam Smith did not trust corporations, but in today's world, in the United States, this is part of our partisan war. On the right you've got people who trust people in business and don't trust government. On the left, you've got people who trust government and don't trust business. Both of them need to be moderated.

I find it quite interesting that when people look at the Tea Party and the Occupy movement there is this clear division: an understanding that the Tea Party is right and the Occupy movement is left. To my mind, they're very similar. Both of them say that the institutions of society are broken. Occupy focused on the corporate sector, Tea Party focused on government. Both of them dislike the connection between the two, and yet they got pulled into the partisan wars, which is really unfortunate because there *is* a broken system.

So, again, I'm trying to change my one piece of it. Am I an outlier? I'm not so sure because I *do* see more and more courses starting to emerge with people beginning to ask the difficult questions that have to be asked about the dominant pedagogy in business schools and how it should change.

The Aspen Institute gives out an award every year called *Ideas Worth Teaching*, and there are some really great courses popping up. One of my favourites is by Rebecca Henderson at the Harvard Business School, who offers a course called *Reimagining Capitalism*. I want to offer that here at Ross too—I think every business school should have that.

I think we should teach business students how to be stewards of the market. And to do that, you don't take the market as given, you don't take capitalism as given. You understand—from Adam Smith to James Madison to the present—that there are multiple forms of capitalism. Scandinavian capitalism is different than American, different than Japanese. Capitalism has changed over time; it's quite malleable. You can't price fix, you can't collude. We now accept those as rules of the game, but they weren't always rules of the game.

So if we start to think in those terms, we can begin to imagine, *"Okay, what kind of capitalism do we have right now? What kind of democracy do we have right now? What kind do we need? How did we get to where we are now? How do we get to where we need to be?"*

I want to teach more business students this. I think there's an appetite for it among the students. I think there's a growing, though small, cadre of business schools starting to ask these questions. Take the issue of income inequality. We need to teach business students about income inequality. I mean, how can you *not* teach business students about income inequality, or climate change? Teach them some social science, teach them some natural science. There are ways to get it into the curriculum, to adjust things so that a more rounded business student is coming out.

I think the conversation is growing. *Will it be stillborn?* I don't know. I hope not. *Will it grow so that it gets to a point where it will really spread and diffuse?* I hope so, but that requires steady, consistent, pressure and innovation. And I'd like to think that I'm part of that force.

Questions for Discussion:

1. To what extent are the political categories of "left" and "right" still mean-ingful? How would you describe them to someone from a completely differ-ent culture?

2. Should every public policy student be forced to take at least one course in the business school and vice-versa? Should all social science students be obliged to take at least one course in the natural sciences?

VII. Opinionated Ignorance

Hardly what the Founding Fathers had in mind

HB: Well, you're certainly doing your best. An important general point to emphasize—which should be obvious, but sadly often isn't—is that going forwards constructively requires actual knowledge of what's happened before. You mentioned Adam Smith and James Madison, and you often discuss the thoughts and writings of the Founding Fathers.

Which brings me to an important point that is hardly limited to business students, but is, with all due respect—which is something people usually say right before they go on the attack—indicative of a real problem with your country: *there is no real understanding of history.*

And there are some deep ironies here, because those very same revered historical figures who are so often trotted out as symbolic puppets for the likes of Fox News in support of their frequently simplistic and ignorant claims—in fact, perhaps Fox News is not even the best example these days with the rise of even more benighted extremists like OAN—invariably said entirely different things than what is regularly ascribed to them.

I had a conversation some years ago with John Dunn, a political theorist from the University of Cambridge who's written extensively on the history of democracy, among many other things.

He maintains that there is a tremendous amount of confusion regarding the origin of the United States of what the Founding Fathers actually believed: in particular, there is this false conflation (you also refer to aspects of this in your book) of capitalism, democracy and the rule of law—three things which are quite distinct—and there is this triumphalist notion which is espoused by all sorts of people on both sides of this tribal political landscape, which is just patently historically false.

AH: Yes.

HB: And incoherent. So how is it possible for you to move forwards in a progressive way to have a meaningful exchange of ideas with people who hold legitimate positions who may just so happen to disagree with you under these circumstances?

It's infuriating for any independent bystander such as myself to be looking at, and it hardly seems to be getting any better.

So, kudos to you for trying to rectify that in your small way, but when you quote Joseph Stiglitz, or your friend at Harvard Business School who talks about reimagining capitalism, it seems to me that it's awfully important that before we talk about the future, we acknowledge and understand where we've *actually been* in the past, rather than living in some sort of cartoon-distorted version of what actually happened. In other words, it's important to be able to openly and truthfully declare, **This** *is what happened,* **this** *is where we got to, and* **this** *is how we might change to get to where we need to go.*

AH: Wow, that was a mouthful.

HB: Yeah—and it didn't even end with a question. That's typical for me.

AH: Well, it's a provocation. It reminds me of a quote I love from John F. Kennedy. He said, "*You can't have the comfort of opinion without the discomfort of thought.*" I feel like we live in a time right now where people aren't thinking, they're just parroting opinion that's coming from their particular tribe: if I'm on the right and Tucker Carlson said it, then I agree with it. And if I'm on the left and Al Gore said it, then I agree with it.

HB: Speaking of which, I have to admit that I was astounded that you managed to respectfully quote Tucker Carlson in *Management as a*

Calling. I mean, citing Tucker Carlson—if that's not sending out an olive branch, I don't know what is.

AH: Well, I agree with him entirely when he said, *"Anyone who worships the market is crazy—the market is a tool, like a toaster oven; and the moment it no longer serves your needs, throw it out and get a new product. Any economic system that weakens or destroys families is not worth having."*

I agree with that 100%. There are other things he says that I don't agree with, but that's part of my overall point. You should be able to say, *"I agree with this person on this point,"* rather than simply saying, *"If Tucker Carlson said it, I don't agree with it,"* just like people on the other side are saying, *"If a professor said it, I don't agree with it."* That's the challenge.

So, how do you change a culture that is so dumbed down that any news on a particular topic will be summed up in a 240-character tweet. I mean, you actually have to read something. You actually have to look at something besides Monday Night Football. I struggle with what you're describing because not enough people in this country are thinking and they fall back on convenient, simple metaphors.

One revealing statistic I've heard—I'll probably get the exact number wrong but it's somewhere in this ballpark—is that only something like 42% of Americans have a passport and that many Americans will never leave this country. Meanwhile they unflinchingly maintain that everyone wants to be American because we are the superior society.

This sort of thing shows up in strange ways. I remember when the United States went into Iraq and you heard metaphors like "Saddam Hussein is like the bully in the sandbox, and we're going to go in and kick him in the teeth". Now you'd think that appreciating the nuances of global politics and unleashing the American military requires a more sophisticated notion than just kicking a bully in the sandbox, but that worked.

And that should scare people: that there is a country with such a large global influence where the level of thought is not what it should be. Again, I go right back to the RAND Corporation report: we don't respect previously trusted sources of information. People are questioning expertise. They wouldn't do it with the quarterback of their favourite football team, but they will do it with other things.

I remember talking to a friend, saying, *"You know, Joseph Stiglitz is really worried about income inequality,"* and my friend replied, *"He's entitled to his opinion."* And I said, *"He's got a* **Nobel Prize in Economics***!"* but he just said, *"Well, I don't see it the same way."* And I was just—I mean, my jaw was on the floor. That is a serious problem: we have denigrated expertise.

HB: Absolutely. I'm looking forward to kicking your fellow Americans even harder in the teeth later on. But before I do, I want to say something positive—

AH: And everyone in France is totally respecting of—

HB: That's a cheap shot, Andy: by kicking your countrymen in the teeth I'm hardly saying that the French are perfect, because they're certainly not, but you'll have to wait until I get there and I hope you can mount a better defense than that. At any rate, I want to say something positive—an important point related to this topic that I think is far too often overlooked.

So we've alluded to the disconnect between what the likes of Adam Smith and the Founding Fathers actually said and believed and what is now often attributed to them—like saying that they were all stern advocates of democracy and all of that, which is just nonsense.

But one thing they clearly *did* believe in, consistently and almost to a man, was education—which they not only wrote and talked about, they actually *did* something about. If you look at Benjamin Franklin with the University of Pennsylvania or Thomas Jefferson with the University of Virginia, it's very clear that they believed that the notion of a "more perfect union" was predicated upon a more educated and knowledgeable populace.

That was a very, very common belief, with its roots firmly in an Enlightenment tradition that played such a formative role in their worldviews. And that's what I—and many others—naturally associate with the United States of America in terms of its founding ideals: it's a wonderful and important thing.

So when I see that being dragged through the mire—when people are not just ignorant, but *wilfully* ignorant: hostile to anybody who represents any form of education and has put the time in to learn things, as you just eloquently described with *"Joseph Stiglitz has his opinion and I have my opinion"*, which strikes me as just another way of saying, *"Oh, don't trust those elites; don't trust anybody with an education; don't trust anybody who's actually gone through the time to educate themselves, to become knowledgeable about something"*—**that**, to me represents an enormous transgression of those vital values that the country was founded on.

AH: Let's go back to where I started. So yes, we have this problem in this country. How much of it is self-inflicted by academics, those experts who choose not to go out into the public? I think a lot of it is self-inflicted. And I think that we can change our own institutions to try to bring education to many more people. That's the promise, once again, of the new technology, like making MOOCs—Massive Open Online Courses—available to people around the world. We can change the level of education in this country through doing this. Let's broaden the idea of the classroom. Right now, students pay tuition to come into my classroom, but I think about "my classroom" as being much broader: how can I bring what I do, what I write about, what I do research on, into a broader public? By thinking in that way, perhaps we can bring universities into the 21st century to address the problems that our society faces.

Questions for Discussion:

1. What do you think Howard means, exactly, by "this false modern-day conflation of capitalism, democracy and the rule of law"? Readers interested in more details on this point are referred to Chapter 3 of **Democracy: Clarifying the Muddle** with University of Cambridge political theorist John Dunn.

2. Why do you think Andy brings up the fact that only 30-40% of Americans have a passport? How is this relevant to the discussion in this chapter?

VIII. Qualified Optimism

Hopeful signs

HB: Let's move on to more of an optimistic topic and talk about how developments in the business world have the potential to meet some of these existential challenges that we alluded to earlier.

In *Flourishing*, the book you co-wrote with John Ehrenfeld that served as the basis of our first discussion seven years ago, you talk about how societal values need to change and speculate on how the business world might, broadly speaking, be an engine of this change.

In *Management as a Calling*, meanwhile, you go into considerably more detail, describing two phases: an initial "enterprise integration phase" of recognizing the problem, reducing unsustainability, starting to become aware and making some shifts within the climate of business, followed by a "market transformation phase" when systemic structural changes in the business world begin to occur.

So I'd like you to speak to that and give me some reason for optimism, because we've been fairly pessimistic over the past 20 minutes or so.

AH: This is something that has changed considerably since our last conversation seven years ago. When I first started teaching this, it was in the mid-1990s and no other business schools taught it—I was actually introduced at a conference recently as a "grandfather of the field", which I find funny—that shows you how young the field of sustainability in business schools is, or at that time "environmental issues".

So the way to make it legitimate was to fit it in with the business logic, make it fit in with how to make money—reframing climate change in terms of operational efficiency, cost of capital, consumer demand, and so forth. Businesses know how to handle this, and they will address it, as best they can, under those parameters. Now that's good. That's nice. We get things like electric cars from that process. We get innovations in sustainable food, or sustainable hotels, or sustainable clothing. And that's great.

But at the end of the day, the answer lies in a different direction. For example, while the Tesla is a great car, the answer is not another car. The answer is rethinking mobility. So that brings me to where we need to go next: these efforts in enterprise integration are great, but they're not fixing the root of the problem. They're only slowing down the velocity at which we're running into a brick wall, but that's hardly the same as reversing course.

To reverse course, we have to change the system. Now that's a big statement that a lot of people—including me—are throwing around now without a lot of specificity, because we don't yet have clarity on how we change the system.

But there *are* certain clear things that we should change and are starting to change, such as questioning certain metrics within business like gross domestic product and discount rates. These lead us into some pretty bad directions: the standard discount rate of 10%, straight line method—everything 10 years and beyond is worthless. Is that true? Actually, no, it's not. That's called "the future of your children and your grandchildren".

What about Gross Domestic Product? Well, anytime money moves, GDP goes up. If I start my day eating Krispy Kreme donuts, GDP goes up. If I have a heart attack and go to the hospital, GDP goes up. If I die and get a nice funeral, GDP goes up. Clearly, that's not good.

So, for example, back in 2008 Nicolas Sarkozy commissioned a group of experts, including Joseph Stiglitz, and Amartya Sen to come up with some alternatives to GDP; and they have a really nice report about different metrics that are missed in GDP that we need to attend to.

We're starting to have questions around short-termism that's caused by our focus on the shareholder, and we're beginning to rethink things there too. Paul Polman, when he was CEO of Unilever, eliminated quarterly reports, saying, "*I need people in my company to think more long-term.*"

We're starting to question consumption. The economy right now is based on consumption, but perhaps we can start to disconnect profits from material or energy use.

That's where, for example, the shift of mobility starts to come in. Do I *really* want to own a car, or do I just want the ability to move my body and my stuff around with ease? I look at young people right now, students in my classes—many of them don't want to own a car.

Many don't even have a license, which just blows me away because I got my license the first day I could; I got a car on the first day I could afford one, and I have a couple of classic cars. But when I say, "*I love cars,*" some of my students look at me like I've got three heads: *It's a hunk of steel with wheels and a big debt load around my neck—why would I want that?* In the classic car community people are worried about this: is the value of their collectors items going to start to go down? Well, I'll eat that one—that's just part of cultural change.

So I do think there are changes afoot. I think they're happening in business. I think they're happening in the market. Am I ready to say that the momentum is strong enough that we're definitely on a path towards the solution? No. But am I seeing glimmers of hope that there are changes that are happening, that people are asking the right questions? Yes.

HB: Let me ask you a specific question. Seven years ago—and seven years is not that long; I mean, it's not a tiny amount of time, but it's not that long—one of the things that you specifically highlighted was the energy grid. You said to me, "*The energy grid is a joke in this country,*" and talked about how it's going to be substantially transformed to be much more nimble and flexible. So where are we now with respect to structural changes to the energy grid? Are we on the sort of track that you would have hoped seven years ago?

AH: That's a very interesting question. Just this past summer, something very interesting happened in the energy markets: NextEra Energy just surpassed Exxon Mobil as the energy company with the highest market cap, and two other renewable energy companies have now joined them, eclipsing Exxon Mobil and BP. An interesting transition is in play.

Have we seen enough in terms of changing the grid? Well, first of all, no utility right now is going to build a big base load coal-fired power plant. Those days are over. Fracking has introduced a lot more natural gas in the markets since we last talked, which helped bring the end of coal. Will that stay? I'm not sure. The amount of renewables within the market is growing, but have we started to rethink more microgrids? Not to the extent I would like to see, but Biden just chose Jennifer Granholm to be his Secretary of Energy. She is very hot on this topic. Let's see how that plays out.

Is there an opportunity right now? Yes. Something very interesting just happened: here in Michigan we have Consumers Energy, which was very aggressive on renewables. Their CEO, Patti Poppe, has just moved to becoming the CEO at PG&E (Pacific Gas and Electric) based in San Francisco, one of the biggest utilities in the country and a very beleaguered utility right now—but when's the best time to change an organization? When things are really bad. I think she's going to have a really strong hand to try to shift that utility in a state that's already very welcoming to renewables.

So, I'd say that momentum has been building and there are definitely reasons to be optimistic. Of course, as they say, optimism and a couple of bucks will get you a cup of coffee. It's kind of like fuel cells: people like to say, "*Fuel cells are 20 years out and always will be*", but let's see how it plays out. Is the political will there to kick it over the top and change the production tax credits and renewable tax credits? Will we get a price on carbon? It's hard to say.

HB: You talked earlier about harnessing the crisis that is the pandemic we're currently in. Earlier, you specifically mentioned communications technology: that we would likely not have had a discussion like this nearly as comfortably a year and a half ago as now. What types of change do you see might come about as a result of harnessing this particular crisis? Can you point to specific things and say, "*Well, we've* *learned some lesson here"* or, *"We've been shocked into some kind of awareness over there."* Is there anything in terms of a positive silver lining you can see that, once we emerge from this and people aren't walking around with masks anymore, we might be able to point to as something positive that was able to be productively used for the future?

AH: I'm reminded of a cartoon in *The New Yorker* with someone giving a presentation in a corporate boardroom and everyone else is sitting around the table with their heads in their hands. There's a graph on display that's going downward, and he says, *"The bad news is that, during the crisis, people have figured out what really matters."*

I think of what we're going through right now with COVID—it was a test of our institutions. Some of them held, some of them wavered, some of them broke. Others exposed weaknesses. Look at the food system. The food system has shown to have some very serious weak points that should scare people, and *is* scaring people. We've got to shore up the system, we've got to tighten it.

Our ability to trust science, to come to an agreement that we've got a collective problem that we need to address, was shown to be severely wanting, and we've talked about it quite a bit in this discussion so far. To my mind it is a harbinger of other collective crises we face in the future.

Take climate change. The number of people who believe in the reality of climate change in this country is on a steep upward trend. What's driving that? The vivid events that are hitting people we know, right now—and at times hitting our wallets. Look at the wildfires in California. The insurance payouts for that have not hit the books fully yet—they're going to be very, very high. To my mind, insurance is going to be a prime mover in shifting the market to addressing climate change.

If you want to rebuild a house after the wildfires in California, good luck—you're going to have a very hard time finding insurance. And if you do, it's going to be at a significantly higher cost, with much less coverage. Some companies, like AIG, say that if you want house insurance from us you're going to have to buy into our private fire-fighter company, because if there's a fire we're going to try to protect your asset before we pay for its replacement.

That should cause some shudders for some people who will say, *"Wait a minute, so now only rich people can have that house?"* That's going to start to raise some questions about where the state may step in and try to equalize and balance the insurance market.

HB: Come on, Andy—if you can't trust AIG, who can you trust?

AH: Well, what I'm about to say is going to sound sad, but, many people in this country will trust an insurance company more than they'll trust a climate scientist because an insurance company doesn't have a dog in the fight: they're just looking at the numbers saying, *"I'm sorry, but the costs are going up."*

There are some really compelling graphs from places like Munich Re and Swiss Re showing a steady increase in natural disasters over the last 30 years. And then if you look at the payouts—and this is what scares insurance companies—it's very spiky, very up and down. An insurance company doesn't care if there's a steep upward trend, they can amortize that: they have actuarial tables that can tell how to price the instrument to get the right sort of payback.

But now a lot of insurance companies are throwing away their actuarial data that's more than 10 years old and hiring climate scientists to come in to figure out how to price these instruments. If you want to change people's beliefs on climate change, make it salient. And the best way to make it salient is to put a dollar sign on it.

So people *are* changing now, people are shifting. A lot of companies are no longer investing in new coal, that day is over. I do see changes within some bellwether sectors of the economy, insurance and finance, that speak to the idea that maybe we're in the midst of a shift.

Questions for Discussion:

1. To what extent do you think most economists are too wedded to standard measures like GDP simply because that's what they're most familiar with?

2. What sort of concrete actions do you think politicians can take to expedite the needed changes that Andy mentions in this chapter?

3. What sort of concrete actions do you think individuals can take to expedite the needed changes that Andy mentions in this chapter?

IX. Spreading the Word

Creating new platforms through technology

HB: I'd like to return to the topic of the media and trust in science and bring up a specific point that I've thought about quite a lot and see if you think it's true—and, if so, what can be done about it. And once again it's related, I think, to this question of cultural values.

It's about the notion of avowing one's ignorance. It seems to me that, for somebody who only looks at the world through the filter of the media, spin and politics, admitting that you don't know something is pretty well the worst thing that you can do or say, and opens you up to all sorts of scorn and ridicule.

The most important thing is to have an opinion, preferably a very strong opinion, and preferably with all sorts of spin associated with it: "*We won the debate,*" or "*Crime is on the rise,*" or "*Growth was way up,*" where knowing the actual numbers involved don't seem to make any real difference, and the most important thing is to have a very strong view. That's my sense of the framework—the structure of the entire idiom.

And the point I want to make is that all of that is 180 degrees away from the way things work in the scientific world—or at least the way they're supposed to work. In the scientific world, *I don't know* are the three most important words that you need to be able to clearly formulate in order to eventually know something. In short, you have to be able to identify *what*, exactly, you don't know, in striking contrast to what you feel that you *do* know. Otherwise you simply can't make progress.

I often get the feeling that it's this "culture clash" that lies at the heart of so many issues. Because once some scientist will stand up and point out what she doesn't know or at least what she isn't unequivocally certain of, the media will inevitably cry out, "*You see—she doesn't know, she said so herself. None of these people know anything and they're just not trustworthy!*"

To me, that's a very revealing cultural distinction that is used and abused—often unconsciously—by people who have spent their entire lives only looking at things from this political, media-driven side of things. And it also has dangerous repercussions in another way, giving rise to the erroneous conviction that "there are always two sides to every story".

Because in their world, the Republicans say this and the Democrats say the opposite, and vice-versa. And exactly as you were saying earlier, you wind up in a situation where if person A says something I'm automatically going to either agree or disagree with it depending on which tribe he's associated with.

And so in my view that characteristic, in an over-arching way, is perhaps the most dangerous impediment to really understanding the "scientific approach" for most people—I don't even think it's so much scientific per se, but just a question of a rational, problem-solving, moving-forwards approach.

So my first question—I knew I'd come to a question, eventually—is, *Do you agree with that?* And my second question is, *If so, do you think that it's getting worse?* Because it seems to me that, while it was pretty bad the last time we spoke, it seems like it's become even worse lately. And if that's right, how do you deal with that? How do you overcome that?

AH: I think it *is* getting worse, but again, I think it's self-inflicted in many ways. You talk about how all scientists start out with saying, "*I don't know.*" But when they go out into the public, they don't do that.

I can tell you that here, in Ann Arbor, there are many people not connected with the university who are hesitant to invite a professor over for dinner because they don't want to get lectured at. Every time I hear a professor begin a statement by saying, "*The literature says...*" I flinch, because that's a way of putting someone into a corner: if you dare to open your mouth and offer a different opinion, now you're challenging "the literature". I find those kinds of statements unhelpful: they put people in a corner. So that's the first thing I would say.

Beyond that, people in Ann Arbor often meet academics because we've got a major university in our midst. But if you look at things more generally, it's a different situation. According to a relatively recent survey, most Americans could not name a single living scientist, and half of those who could named Stephen Hawking—this was back when he was alive, obviously—because he's in movies, because he's famous. That's a problem. Most Americans are not scientifically literate. Not only do they not know what science says, they don't know how science comes up with its conclusions. They don't know how it works.

To my mind, the challenge for scholars is to not only communicate the results of their work, but communicate *how* those results were derived. A lot of scientists are afraid to do that—it's perceived as sort of "opening up the books", and they become defensive.

HB: Maybe it's a chicken and egg thing, though. As I was saying earlier, maybe a scientist is afraid that if she publicly admits what she doesn't know, she'll be pilloried and mocked for her "ignorance" by a combative media. Do you think that plays a role?

AH: That's true. Going back to Roger Pielke's book that I mentioned earlier, *The Honest Broker*, he had a very provocative conclusion that stirred a lot of debate within scientific circles. He described four roles, and two of them I think are relevant for what we're talking about now: "the issue advocate" and "the honest broker". The honest broker takes all the science and puts it out to the public. The issue advocate narrows it down and says, *"This is what you need to know."*

For example, we don't take all the climate science there is and throw it to the public. The IPCC (Intergovernmental Panel on Climate Change) narrows it down and comes up with a consensus statement: *This is what the consensus of the literature says.* He called that the actions of an issue advocate. And then he said that what we should be is honest brokers and throw it all out there.

I leave that as an open question, because I think a lot of people would be very uncomfortable just throwing all the science out there and letting people make sense out of it, because it's very convoluted, it's very complex. I can guarantee most people have not read the IPCC reports, much less all the science that's out there, so to expect them to make sense out of it is really difficult.

Many people *are* defensive of science: they're dealing with a critical and hostile audience that is looking for weaknesses to pick at.

And then there are these language issues: in scientific terms, "uncertainty" is an error term, while in the public it means, "*Oh, you don't know.*" There's a language barrier here, so it's very difficult. This is not something that is simple: just go out and start communicating your work. It requires some training and some understanding of the language you use, the language others hear, and the disconnect between the two.

HB: Well, I would submit to you that there's a specific connection that could and should be made here to a central theme in *The Engaged Scholar*, namely the role that the university can play in bridging this gap.

I'm not a climate scientist, obviously, but were I to be one, I would be deeply reticent to go on Fox news, say. It's not even a question of left or right, it would be almost as worrying going on CNN or MSNBC and being forced to justify a sophisticated scientific conclusion in 60 seconds while being peppered with questions about my political leanings.

I'm not disagreeing with you that more can be done, but I do think that it's worth mentioning that there is a significant lack of opportunity for knowledgeable individuals to coherently and thoughtfully present their cases.

You graciously cited Ideas Roadshow in *The Engaged Scholar*, as well as other new initiatives dedicated to getting scholarly ideas out into the public domain. But one of the things that has always mystified me is why universities don't do much more than they do.

As you know, in a previous life—likely due to some sort of hein-ous sins I'd committed in a still previous life—I found myself an academic administrator. And when you're an academic administra-tor, you naturally find yourself talking to all sorts of other academic administrators—because misery loves company—and a very, very common complaint that I heard from university presidents or provosts or whatever was something like,*"We have all of these great faculty at our university and they're doing all these fantastic things, but nobody knows about them."*

It struck me at the time as a very strange thing to say—and I have become even more convinced by its strangeness since—because the implication seems to be, *"These guys are so busy and will never make time for anyone to talk to them,"* or *"They have no communication skills and are unable to convey to a non-expert what they are doing."*

And that's complete nonsense. It was very clear to me then, and it's even clearer now, that if you simply make a reasonable effort to provide an appropriate forum for researchers to talk openly and honestly about their views—the things they are most proud of, the things they are most frustrated by, the things they thought they understood but later turned out not to, the things they still don't understand—then virtually all of them will jump at the opportunity to do so. As you know, we've held over 100 such conversations for Ideas Roadshow now and by and large the biggest problem I have is getting people to *stop* talking about their work and keep things down to under 3 hours or so. Which is pretty well the exact oppos-ite of what all these university administrators were saying and are doubtless still saying.

So one of the things that completely amazes me is why univer-sities, virtually all of which have access to vastly more digital media equipment than you'd ever need, along with all sorts of other pressing needs that might be easily met through the use of this technology—like inspiring alumni with all the great things that their researchers are doing—simply don't bother doing anything with this equipment other than occasionally sticking a camera up in a classroom and recording a lecture. Why don't they use the technology to provide an appropriate forum for their research staff to explain what they're doing to a broad general audience? Because I can assure you from personal experience that most active researchers would very much welcome such an opportunity.

AH: I see a number of points to draw out from what you just said. Personally, I've often talked to academics and suggested something like, "Why don't you write for *The Conversation?*" And a frequent response I get is, "*I don't have the time.*" And I just roll my eyes. I mean, here I am talking to you, which is a three-hour investment of my time. I can certainly afford three hours.

HB: And it might well turn out to be the best three hours of your life.

AH: I mean, if you write a paper that takes you over a year and a half to produce, are you telling me that you don't have an afternoon to sum that paper up in a 750-word essay? I think that's nonsense. If you can't do that, I don't think you understood what you were working on all this time. I don't believe that.

And going beyond that, we also have to consider how technology impacts the moment of change that we're in right now. If you had asked me to do what we're doing right now a year ago, I wouldn't know how to do it. Right now, I'm sitting in front of a camera with two tech guys in front of me from our media lab with all sorts of bells and whistles that many faculty are now using to teach online with. The technology was there, but COVID is now making it more relevant.

These tech guys have a solid career future in front of them because what they are providing universities are going to want more and more of and we're now seeing the value of it.

For example, I had Paul Polman speak in my class through social media, and it was great. There's a whole new lexicon that has emerged as the university has gone into this virtual world: we now talk about synchronous content and asynchronous content, which I find quite amusing because asynchronous content used to be a book.

Now we do video. But, if I'm doing video, what kind of video am I doing? I know some colleagues who make asynchronous content by basically just taping themselves over a PowerPoint deck. If they think that's going to work for the tuition-paying public, they're out of their mind.

So what would constitute quality asynchronous content? Well, not to be self promoting, but maybe what we're doing right now would be valuable asynchronous content, or maybe the work of someone like Ken Burns. If I was taking a course on the Civil War, would I prefer to hear a professor talking over a PowerPoint deck or watch a Ken Burns episode? Which one will I learn more from?

The new faculty coming into this academic world are going to have to be able to have skills, and be adept at knowing this technology and how it works, and how to work with these guys to really produce dynamic, engaging content. I'm actually glad I'm coming to the sunset of my career rather than starting my career, because I have to admit that this scares me, but I'm willing to bet the Young Turks coming in right now are thrilled at the possibility of what this medium allows us to achieve.

HB: Let me say a few words about the fear factor, because I think that's important to address. As somebody who's been in the "digital media scholarly content" game now for eight years or so, to me the really interesting point is that this technology enables entirely new and important sorts of content to be created in a high-quality and relatively painless way.

As you know well, all I say to people is, *"Give me two hours or three hours of your time. You don't have to prepare. You don't have to write anything down. We can just start having a conversation. After all, it's about your work and your career and your life, so there's no need for you to have to prepare for that."*

AH: There's no other part of my professional life that is as unstructured as what we're doing right now. When I walk into a seminar room, I got my PowerPoint deck, I know the rules. What you and I are doing right now, on the other hand, is different. I have to admit, I had some butterflies walking in here—I don't know what's going to happen. I lose control. Personally, I'm okay with that. But it's very different.

I have a colleague who went on the *Bill Maher Show*; and I said, "Really? Bill Maher?" And he told me, "*Andy, I'll go on Rush Limbaugh, I'll go on Howard Stern. I'll go anywhere I need to go to bring my knowledge of climate science to the public.*" Good for him.

I was just reading an article about Obama's Chief Economic Advisor, Austan Goolsbee. He's at the University of Chicago now and he goes on Sean Hannity's Fox News show. Hannity introduces him as "a guy that destroyed the world" and mocks him. But they have a debate; and he says that he does it because if he doesn't people aren't going to get quality content. Good for him too. I don't know if I'd have the stomach to do it, but good for him. He actually responds to hostile email.

I have a separate folder for hate mail—I keep it in a folder and don't respond. Everyone has to figure out how much of a stomach they have, or how far they want to immerse themselves into the tensions and the challenges and the hostility of this kind of engagement.

HB: So I applaud those people, unequivocally. I respect what they're doing and the sacrifices they're making, because I'm sure it's not fun to be in that situation. But my point is that there are ways around it.

In particular, there are things that the university could—and in my judgment, *should*—be doing, using this technology. And when I say "this technology" I should probably be more specific. When I started Ideas Roadshow the only thing I was focused on was the cameras, and maybe the lights. My thinking was, "*I should be able to do something truly innovative and worthwhile by capitalizing on these new portable, high-quality cameras and go around filming people*".

Well, that's all true, but this goes back to what you said before about anxiety and a new world and all the rest of that—what I've come to appreciate is that, in fact, the transformative technology is actually not so much the cameras. Obviously they're necessary, but these days the truly transformative technology is actually the editing software, like what these guys at Adobe have managed to do with things like Premiere Pro and their creative suite.

So the point here is that if you are somebody who is motivated to create original, high-quality content, you can actually now do so virtually independently. It's really quite amazing. And universities should be doing that: they should be using these tools to capture and transmit the ideas of their faculty.

AH: And maybe they will. I'm willing to bet that the equipment that I'm sitting in front of right now did not exist in this building, maybe not even on this campus—definitely 10 years ago and maybe even 5 years ago. And I'm willing to bet that the job description of these two gentlemen in front of me definitely didn't exist 10 years ago, and maybe not 5 years ago; I do think COVID is going to change the game.

Questions for Discussion:

1. *Do you agree or disagree with Howard's remarks of a difference in attitude between those coming at issues from a media background as opposed to an academic one?*

2. *To what extent has modern technology impacted the world of ideas-related content? Are there as many engaging opportunities to learn about research and scholarship as you would like?*

3. *Is it reasonable to expect that news and comedy shows should be the primary sources of research and scholarly content to the general public?*

X. Getting Personal

Slings, arrows, and the learning experience

HB: Before I move on to my polemic remarks that I've promised to harangue you with, I'm going to ask a more personal question. You mentioned just now, in an off-handed way, about receiving hate mail and that sort of thing. You're obviously not somebody who whines and complaints a lot, but I'm guessing that there must have been times when the reactions you were getting were difficult to absorb personally. Are you getting any better at dealing with that? Is that something that you think about a lot or does it take a toll on you?

AH: Well, I'm getting better at handling it; and the way I'm doing so is that I think of what I'm doing as learning. So what lessons can I learn from my experiences with encountering vitriol? What can you learn after being publicly confronted with someone holding a Bible and telling you, *"The seas can't be rising because God promised that He would never flood the earth again."*? That was a pretty tense moment.

When I first got hate mail, I wouldn't say I was catatonic, but I was really thrown back. Now I just take it as part of the game, and mine is pretty mild. Take a look at Katharine Hayhoe at Texas Tech or Michael Mann at Penn State; Michael Mann even had white powder mailed to him—some pretty awful stuff.

So mine is mild compared to that, but it still stings. Let me give you an anecdote. One of my first papers on climate change and culture caught the sights of Mark Morano, who used to work for Rush Limbaugh and James Inhofe. He went at me very aggressively and posted my email address. That's when the hate mail began. It caused a brouhaha in social media.

An important thing to remember about social media is that it never goes away—that legacy is there. If you Google me, you can find it. It was all directed at me. I didn't engage it.

Anyway, about four or five years ago, I got a call from Fred Krupp at the Environmental Defense Fund and George Schultz, former Secretary of State, asking me if I would facilitate a conversation among the CEOs of natural gas companies to voluntarily reduce the fugitive emissions of methane.

I was thrilled: what an important conversation and *I* could be part of it! About a week or two before that meeting, one of the CEOs looked me up—they were a pretty skittish group to begin with—saw this brouhaha and said, "*Hoffman's a radical. I don't want him in the room.*" And I was out—just like that.

I had a long conversation about it with a colleague, a respected third-party mediator, John Ehrmann. And he said to me, "*Andy, I've developed my reputation as a third-party mediator by having no public position on anything. You're a professor, this is what you do. Would you have done things differently if you knew this outcome?*"

And I replied, "*No, but I would at least like to have known the possible ramifications for doing it.*"

So there's my opportunity to turn that experience into a lesson for others. I'm not saying "*change your messaging*", but rather, "*Remember that things are permanent.*"

This also speaks to the number of tweets I see out there. I think of the professor in African-American studies who tweeted that the problem with the world is that it has too many white men in it; and I wonder, *Really, was there an upside to doing that?* I don't see any upside. And the downside was huge.

So it's important to just recognize the cautionary tales of social media, recognize the complexities of the Wild West we're stepping into. So, yes, it's been hard at times, but all learning is hard. All learning is painful. If it wasn't painful, I don't think you've learned.

That sounds very glib. I don't mean it to be, but I was clear from the start why I became a professor. I won't go into the details of it, but I sort of backed into an academic career. If you had told me 30 years ago that I'd be a professor sitting here talking to you right now, I would have laughed and said, "*You've got me mixed up with somebody else*." And I think that has enabled me to have a certain ambivalence, which has been extremely healthy.

If I didn't get tenure, I can honestly say that, while I clearly would have been bothered, I would have rolled with it and moved on. I know people who've been denied tenure and their life just came to a grinding halt, as well as people waiting for tenure who just became totally fixated on that result. I find that extremely unhealthy.

HB: Well, in my mind this is all of a piece with what we were saying earlier about your values and your motivations.

AH: Yes.

Questions for Discussion:

1. Would you support measures to trace personal attacks through email and social media at the expense of protecting your privacy?

2. To what extent does the threat of receiving abusive comments from radicals impact the willingness of academics to engage with the general public?

XI. Shattered Leadership?

Desperately trying to pick up the pieces

HB: OK, so now it's finally time to move on to my cataclysmic conclusion that I've threatened you with several times already.

You describe, in *Management as a Calling* and elsewhere, how climate change, in addition to a spectrum of other issues characteristic of "life in the Anthropocene", is "the ultimate commons problem"—an undeniably international problem that naturally calls for coherent international measures.

These issues, after all, represent existential threats to humanity; and in your writings, you frequently highlight the often-overlooked point that however difficult life is for us when we have to grapple with the consequences of climate change, they take a far greater toll on poorer people and poorer countries. This is well accepted, and you see it in this microcosm with the pandemic both in terms of the struggles of poorer countries and the struggles of poor and underprivileged areas within richer countries. After all, it's hardly surprising that those who have fewer tools and fewer resources would have a harder time adjusting to the impact of a changing environment.

So these Anthropocene, "ultimate commons" problems necessarily call for a coordinated international response, but what seems undeniable to me and obviously important to mention when you compare the situation that we're in today with that of our first discussion seven years ago is that things are much, much worse because there is now a real crisis of leadership. And there is a real crisis of leadership because America is no longer seen to be a credible international leader. And this is a real problem.

And before I continue any further, I should say for the record that I'm hardly a card-carrying member of the Democratic party and this is not about politics per se. But I honestly feel that I would be remiss in having a discussion with you about what to do about climate change if we didn't explicitly mention the fact that the United States is no longer in the position of global leadership that it once was—to put it very mildly.

I think this is a pretty obvious point to people outside of the United States, but those within it seem to miss it somehow—perhaps understandably, I don't know. But the important point to make here is that what I'm talking about does not boil down to *"There was some specific politician that I don't like or didn't agree with"*, which is often the way it's portrayed.

So let me try to be more specific. Obviously it's hardly my place to speak for "the rest of the world", but let me just try to give you a candid sense of how it looks from where I sit.

First of all, the 2015 Paris Accords were *a really big deal* for a lot of people—a really big deal because after years and years of foot-dragging and all sorts of empty rhetoric, for the first time there was a genuine sense of optimism throughout the world that, *Gosh, maybe something meaningful could actually be done*. Whether or not the targets were absolutely perfect or were likely to be met in all instances paled in comparison to the symbolic nature of the result that so many countries around the world could get together, despite all the structural impediments and difficulties, and begin to lay a genuinely substantial groundwork for a better future. Over the years, I've spoken to a large number of people around the world, from scientists to bureaucrats to businesspeople, who were incredibly heartened by this.

So when the United States government not only abrogated its responsibility and summarily announced it would withdraw from the accord, that was a major, major blow to a great many people around the world.

But it's not just that. You had a government that is single-mindedly dedicated to doing everything in its power to increasing wealth inequality and victimizing the poor, you have a president who is stridently indifferent to global needs while callously labelling places "shithole countries" and—well, I could go on and on, but there's no point in doing so.

Because—and this is what most Americans seem to be missing—that's not the worst part. By far the worst part in terms of the future of American leadership is that despite all of that, over **74 million people** voted for the guy.

So if you're Angela Merkel or Emmanuel Macron or Justin Trudeau or any other reasonable, democratic leader, the clear implication of all of this seems to be: *We can no longer trust America.*

In other words, it's not just, *We prefer one guy over another, and thank goodness the guy we like is now on his way in.* It's much, much worse than that. The sentiment is much more, *OK, adults seem to be coming into the room for the moment, but who knows what's going to happen four years from now? Anything that we agree to now will likely be unhesitatingly torn up when the next megalomaniacal demagogue comes along.*

Like so much of what we've been talking about today, it's a question of values. And that essential, basic level of trust that the American people share core values with most of the rest of us has now been shattered, which makes the notion of an "American leadership" pretty problematic, if not downright oxymoronic.

Rest assured that despite my kidding around earlier I'm not saying this with any degree of satisfaction whatsoever: I am appalled by it. But my sense is that most Americans don't appreciate what has really happened—for them it's all about whether there's a Biden administration or a Trump administration or whatever, and in my view it's much, much deeper than that: America is no longer seen as a place that shares essential global values and can no longer be trusted to act responsibly in our collective interest—which, to get back to my original point, makes it awfully hard to coherently address, let alone solve, an "ultimate commons problem" such as climate change.

So obviously this isn't your fault, but from where I sit it is the elephant in the room and needs to be addressed, which is why I brought it up. So how would you respond to all of that?

AH: I would respond by saying there are many Americans asking the exact same questions you are. We're trying to make sense out of the vote. We're trying to make sense out of the last four years. What does it mean? And where would we go from here? We can't go back to what we were four years ago; we're going to become something different.

Will there be more distributed leadership around the world? Possibly. Is that necessarily a bad thing? I'm not so sure. The US position...you used the word "shattered". I'm not so sure about that. I think world leaders are pragmatic people, and they will be a little more circumspect now than they were before, recognizing that there is a large number of Americans who would gladly turn their back on everything that was agreed to by the prior administration.

David Brooks of *The New York Times* talks about this, predicting how we're going to have this pendulum swinging back and forth with every new administration undoing what had been done by the previous one. I hope that's not the case, because if it is then we should be very afraid of anyone ever trusting us again. Who's going to do a business deal with someone knowing that the next CEO in four years is going to totally undo every deal we just made?

It might also be that we just got a really good look at some elements of who we are that we may not like. And we will adjust. We got into this mode of "America first" and balkanization that empowered other leaders around the world to adopt a similar stand, attacking the media as fake news and so forth. Maybe we got a good look at that and we'll recoil and say, *That's really not the direction we want to go in.* Maybe there will be a reckoning within the Republican party to say, *Wait a minute, Trump does not represent who we are*, and there will be an adjustment there. The future is in flux, but I can't disagree with anything you just said. I can only add to it that many Americans are asking the exact same question.

HB: But not enough, I think.

AH: Well, I'm not sure about that. I think that there are a lot of people that were horrified by the events of the past two to three weeks, probably not just on the left, but on the right as well.

HB: Well, let's talk about that, because that's by far the worst of all, right? I mean, let's be candid now. I think it was a travesty and a tragedy to pull out of the Paris Accords, but the Paris Accords are chicken feed compared to the idea of a completely unsubstantiated, blatantly partisan attempt to override the democratic will of the people that was actively supported by *many* key members of a major political party.

OK, I'm getting angry here, and obviously this isn't your fault, but this is just a *fact*: *right now* there are over **100 congressmen** who are actively supporting the attempt to subvert their democracy. If this does not make the United States look like a banana republic, I don't know what does. That, too, hardly reinforces your global leadership credentials.

AH: I know there are many Americans who are stunned and horrified by what has happened. We are a crippled nation right now, politically.

HB: So how can we realistically get beyond this? Frankly, it's pretty hard for me to see how. Because I appreciate that, against my best intentions, this might all sound very triumphalist and smug. OK, full disclosure: I'm a Canadian, but I'm not one of these self-congratulatory Canadians who enjoys looking sneeringly and disdainfully at the United States—and there *are* many of those, by the way. But not me.

I believe that, despite its obvious faults, the United States has been a remarkable force for global progress in many different ways. The last time we met I gave you the standard bemused litany of good and bad: here's a place that seems patently unable to engage in the most basic sorts of societal measures that every other functioning first-world country has long sorted out—like ensuring that all citizens have access to basic health care and are denied access to machine guns—while at the same time being the unquestioned global leader of just about any area of research and scholarship you'd care to mention.

But that was then. This is the horse of a completely different colour: what we're talking about now isn't just, "*Oh, they're different, but they have strengths and weaknesses*".

Yet again, it seems, we're talking about values. This is nothing less than a profound and obvious subversion of what every American likes to loudly pronounce is a core value of the country. And I simply can't understand why this is tolerated *by one person* let alone actually **endorsed** by over **100 congressmen**. That's unconscionable.

AH: Honestly, I can't make sense out of it for you right now. I am as stunned as you are; and everything you said I could have easily said myself. I don't know. I honestly don't know how to fix this. If I could fix it, I would go back to the idea of how we come down to a common set of facts. How do we constrain information, call out real fake news for what it is? Maybe there will be some movement now and social media companies will start taking more responsibility in curating their content.

The idea that you now have Twitter adding a tag to every tweet by the president of United States to say, *What he just said isn't true,* is actually quite astonishing. I remember at the beginning of the Trump administration, people within *The New York Times* were having really painful discussions in private rooms, asking, "*Can we actually use the word 'lie'?*" They wouldn't use it at the beginning, but now the gloves are off: they call them lies.

So I think we all got stunned. I don't think someone can necessarily repeat what Donald Trump did because it was so extreme that it was just clickbait all the time—but clickbait works right now: if you want to become famous, say extreme things—extreme things pass through social media faster than more common truths, less sensational truths. I don't have simple answers to your questions.

HB: Well, of course I'm not asking you to have any solutions. And as I said before, it's absurd to hold you accountable for any of this in any way, I just think it needs to be said because, as I said before, it's the elephant in the room right now so it needs to be said.

I also appreciate that you have all sorts of constraints. You have to be much more circumspect because you're trying to move the ship forward and educate your fellow citizens and encourage other people to do something similar. And that's a very laudable goal.

AH: And in keeping with what you said earlier, I also, to some extent, want to stay in my lane. So, to feed this back into our earlier discussion, what can we do about it? Well, as purveyors of knowledge, the university can play a much stronger role in communicating that knowledge to the regular public and politicians. Of course there's going to be some blowback: *You're just a bunch of liberal, latte-drinking professors!*

Well, if that's the worst you can throw out at me, I think I can weather that storm in order to bring important work into the general public to serve our society and to serve the world. Because the dominant influence the United States has on the world can't be ignored. And the idea of balkanization is something we can't afford either: we're at 7.5 billion people, soon to be 10 billion people.

These are collective problems. COVID was a really important test to say, *This doesn't stop at national borders.* The idea of closing the border between Canada and the United States—what's that going to do? That's not going to do anything. The virus is moving. We have to formulate a collective response to this problem, just like we have to formulate a collective response to climate change, to species extinction, ocean acidification, income inequality and so forth.

You talked about poor countries and poorer regions suffering disproportionately from COVID, it's also happening with the vaccines. Who's getting the vaccines? Who's not? Those who can't afford it. Nobody with any kind of basic sense of humanity can look at that situation and say, "*That's cool.*"

HB: Indeed. Anything we haven't talked about that you wanted to touch on? Anything you'd like to add?

AH: I don't think so. I think you really hit on the major elements of what I've been doing in the last seven years in terms of rethinking the role of the university in the academic and society and rethinking business education.

It's funny, now that these two books are coming out soon, I'm feeling something that I've never heard an academic say before: do you ever get to the point where you think, *I've got nothing more to say*? We always have another study. We always have something to say. But right now I feel like I don't know what else to say. I can't come up with something.

HB: It's almost Christmas, Andy. Just take some time and relax a little bit. That's what you need to do.

AH: That's all I'm about to do, Howard.

HB: Good. Well, it was a real pleasure talking to you as always. I've enjoyed it very much. I don't think we can have any more Ideas Roadshow conversations, because we've already broken precedent by having a second one, but I hope that we can have some more face-to-face off-camera conversations—or at least what passes for such these days.

AH: Actually, I'd like to continue the conversation of how to get universities to do more of what we're doing right now, using the technology that's sitting in front of me right now to fulfill the role that I think we should be playing of bringing knowledge and facts and information to both the general public and politicians.

I think we have an opportunity to do that, and we have an obligation to do that. I would like more people to explore what's possible here and invest intelligently in those new possibilities. I think our classrooms need to be totally rebuilt to include technology to allow what we're doing right now to be more a part of what we do in our classes.

I taught my courses in hybrid format this semester. I've got a big plasma TV propped up on a stand. I've got two screens in front of me, two screens behind me. I'm like the Wizard of Oz. That is a total Rube Goldberg. All of this has to be made seamless where social media can't be confined to a media lab. A special room has to be integrated into our classroom in a way that allows us to use all the tools at our fingertips so as to educate those that walk into the classroom while also bringing our education to the broader public. So there's an opportunity right now. And I would love to further that agenda.

HB: I'd be very happy to have that conversation. The one thing I would say right now is that, sure there's all this technology—and I'm sure you're right that it has to be better streamlined and better integrated and better harnessed to incorporate in a teaching environment.

But in my judgment, the biggest overlooked asset that a university has in order to maximize the power of current technologies is its people: it's not capitalizing on its people. You have undergraduates, you have graduate students, you have postdocs—all of whom could be involved in meaningful and innovative content development in all sorts of important and stimulating ways. They would love it. They would learn something from it. And it would effectively cost nothing to produce, simply harnessing what has already been invested.

Anyway I'm happy to have that conversation off-line if you think it would be useful. I hope we get a chance to see each other again, but I'd like to take a moment to thank you once again for all your time and effort. It's been a real pleasure talking to you.

AH: You're very welcome. And I don't mean this just to be gratuitous, but I think that what you're doing with Ideas Roadshow is ahead of the curve, and I'm hoping the curve catches up to you. I hope what you're doing right now really starts to take off.

HB: Thanks very much. Me too.

Questions for Discussion:

1. Do you agree that America has lost its position of international leadership? If so, how specifically do you think that will impact global efforts to tackle climate change and other pressing issues?

2. What role did the media play in the American political crisis that resulted in the Capitol riots of January 2021? Might this motivate universities to be more socially engaged purveyors of information?

Continuing the Conversation

This conversation was based upon two of Andy's 2021 books, *The Engaged Scholar* and *Management as a Calling*.

Ideas Roadshow Collections

Each Ideas Roadshow collection offers five stimulating expert conversations presented in an accessible and engaging format.

- *Conversations About Anthropology & Sociology*
- *Conversations About Astrophysics & Cosmology*
- *Conversations About Biology*
- *Conversations About History, Volume 1*
- *Conversations About History, Volume 2*
- *Conversations About History, Volume 3*
- *Conversations About Language & Culture*
- *Conversations About Law*
- *Conversations About Neuroscience*
- *Conversations About Philosophy, Volume 1*
- *Conversations About Philosophy, Volume 2*
- *Conversations About Physics, Volume 1*
- *Conversations About Physics, Volume 2*
- *Conversations About Politics*
- *Conversations About Psychology, Volume 1*
- *Conversations About Psychology, Volume 2*
- *Conversations About Religion*
- *Conversations About Social Psychology*
- *Conversations About The Environment*
- *Conversations About The History of Ideas*

All collections are available as both eBook and paperback.

www.ingramcontent.com/pod-product-compliance
Lightning Source LLC
Chambersburg PA
CBHW030237030426
42336CB00009B/134